Synthesis Lectures on Electromagnetics

Series Editor

Akhlesh Lakhtakia, Department of Engineering Science and Mechanics, Pennsylvania State University, University Park, PA, USA

This series of short books focuses on a wide array of applications on electromagnetics, particularly in relation to design and interactions with advanced materials and devices. Topics include cutting-edge applications in bioengineering and biomaterials, optics, nanotechnology, and metamaterials.

Ashanthi Maxworth

One Hundred Applications of Maxwell's Equations

Ashanthi Maxworth
Department of Engineering
University of Southern Maine
Gorham, ME, USA

ISSN 2691-5448 ISSN 2691-5456 (electronic)
Synthesis Lectures on Electromagnetics
ISBN 978-3-031-73783-1 ISBN 978-3-031-73784-8 (eBook)
https://doi.org/10.1007/978-3-031-73784-8

This Springer imprint is published by the registered company Springer Nature Switzerland AG
The registered company address is: Gewerbestrasse 11, 6330 Cham, Switzerland

If disposing of this product, please recycle the paper.

I dedicate this book to my loving companion Koda!

For everyone who supported me through an action, word, or a thought, thank you!

Contents

Nomenclature

A	Magnetic vector potential (Wbm^{-1})
B	Magnetic flux density (Wbm^{-2})
C/C'	Capacitance/Capacitance per unit length (F/Fm^{-1})
D	Electric flux density (Cm^{-2})
E	Electric field intensity (Vm^{-1})
$F/F_E/F_M$	Force/electric force/magnetic force (N)
G/G'	Conductance/conductance per unit length (S/Sm^{-1})
H	Magnetic field intensity (Am^{-1})
I	Current (A)
J	Current density (Am^{-2})
L/L'	Inductance/inductance per unit length (H/Hm^{-1})
M/M	Magnetization/mutual inductance (Am^{-1}/H)
N	Number of turns
P	Poynting power (W)
Q	Electric charge (C)
$R/R/R'$	Radial vector/resistance/resistance per unit length $(m/\Omega/\Omega m^{-1})$
A, S	Surface area (m^2)
T	Period (s)
$W/W_E/W_M$	Work/electric work/magnetic work (J)
Z/Z'	Impedance/impedance per unit length $(\Omega/\Omega m^{-1})$
a, b, d, h, l	Distance measures (m)
c	Speed of light (ms^{-1})
f	Frequency (Hz)
i	AC (A)
k	Wave number $(radm^{-1})$
m, n	Mode numbers
q	Electric charge (C)
r	Radial distance (m)

t	Time (s)
$\mathbf{v}$	Velocity (ms^{-1})
α	Attenuation constant (m^{-1})
β	Phase constant ($radm^{-1}$)
γ	Propagation constant ($radm^{-1}$)
δ	Skin depth (m^{-1})
ε	Permittivity (Fm^{-1})
η	Intrinsic impedance of a material (Ω)
θ	Zenith angle (rad)
λ	Wavelength (m)
μ	Permeability (Hm^{-1})
ρ	Radial direction in cylindrical coordinates (m)
$\rho_l/\rho_s/\rho_v$	Line charge density/surface charge density/volume charge density ($Cm^{-1}/Cm^{-2}/Cm^{-3}$)
σ	Conductivity (Sm^{-1})
$\phi/\phi_e/\phi_m$	Flux/electric flux/magnetic flux (C)
$\chi/\chi_e/\chi_m$	Susceptibility/electric susceptibility/magnetic susceptibility
ω	Angular velocity ($rads^{-1}$)
Φ	Electric potential (V)

Introduction to Maxwell's Equations

One scientific epoch ended, and another began with
James Clerk Maxwell.

—Albert Einstein

The man who changed the world forever—King's
college archives

In this book, you will learn about 100 applications of Maxwell's equations—or correctly
named Maxwell–Heaviside equations. James Clerk Maxwell (1831–1879) was a Scottish
physicist considered the father of electromagnetism. Electromagnetism indicates the interaction between electric and magnetic fields. Maxwell's equations comprise Gauss's Law
for electric fields, Gauss's Law for magnetic fields, Faraday's Law, and Ampere's Law
with Maxwell's contribution of the displacement current. Maxwell originally expressed
the theory of electromagnetic fields as 20 equations. It was English mathematician and
physicist Oliver Heaviside (1850–1925) who condensed those to the current four equation
form. Hence, in full, we call those Maxwell–Heaviside equations.

The Gauss's Laws are named after the German mathematician Carl Friedrich Gauss
(1777–1855). The Gauss's Law for electric field says the net flux coming out of a closed
surface is proportional to the charge enclosed. The proportionality constant being one,
Gauss's Law says the net flux through a closed surface is equal to the charge enclosed.
In equation form, this is:

$$\phi_e = \oint_s \boldsymbol{D}.\boldsymbol{ds} = Q_{enclosed} \tag{1.1}$$

© The Author(s), under exclusive license to Springer Nature Switzerland AG 2025
A. Maxworth, *One Hundred Applications of Maxwell's Equations*, Synthesis Lectures
on Electromagnetics, https://doi.org/10.1007/978-3-031-73784-8_1

In Eq. 1.1, ϕ_e is the electric flux and D is the electric flux density or the electric flux per unit area.

The Gauss's Law for magnetic fields states the same for magnetic fields. But when it comes to magnetic fields there are no magnetic monopoles. Hence flux coming into a surface must go out of that surface making the net flux through the surface zero. This is expressed as:

$$\oint_s B.ds = 0 \tag{1.2}$$

In Eq. 1.2, B is the magnetic flux density.

Faraday's Law is named after the prominent English scientist Michael Faraday (1791–1867) who initially indicated the interactions of electric and magnetic fields. Towards the end of his life, Faraday sought an apprentice to carry on his work and prove the experimental results he observed mathematically. Fortunately for the world, he met young Clerk Maxwell who was always fascinated by Faraday's work on electromagnetics. Faraday's law links the electric and magnetic fields. In simple terms it says, a time-varying magnetic field can induce an electric potential difference across a closed loop.

$$\oint_l E.dl = -\frac{\partial \phi_m}{\partial t} \tag{1.3}$$

In Eq. 1.3, ϕ_m is the magnetic flux, and E, is the electric field intensity. The integration of the electric field intensity along a path from one point to another gives the electric potential difference between those two points. Therefore, $\oint_l E.dl$ is a potential difference or a voltage. Since this voltage is applying a force on the electrons on a conductor, this is known as the electromotive force. This electromotive force creates a current, which in turn creates a magnetic field. This secondary magnetic field acts so that it opposes the increase of the original magnetic field, hence the negative sign on Faraday's Law. Faraday made this observation, but it was the Russian physicist Heinrich Lenz (1804–1865) who stated the negative sign in the equation. Without that opposing action generated from the electromotive force, the system violates the energy conservation law.

The fourth Maxwell's equation is Ampere's Law named after the French physicist and mathematician André-Marie Ampère (1775–1836). Which states the magnetic field around a closed loop is equal to the current enclosed.

$$\oint_l H.dl = I_{enclosed} \tag{1.4a}$$

In Eq. 1.4a, H is the magnetic field intensity. Although this equation is valid in conductors, it does not capture the case where there is a change in electric flux. Maxwell introduced a component to the Ampere's Law called the displacement current, which completes this equation as:

$$\oint_l H.dl = I_{enclosed} + \frac{d\phi_e}{dt} \qquad (1.4b)$$

In Eq. 1.4b above, ϕ_E is the electric flux. This term is known as the displacement current which plays an important role in electric capacitors and semiconductor devices.

The term displacement current completed the theory of electromagnetic fields. Faraday's Law states that a time-varying magnetic field can create an electric field, and Ampere's Law with the displacement current term states that a time-varying electric field can create a magnetic field—hence the theory of electromagnetism.

Constitutive Relationships

Let's introduce some of the inter-relationships between the parameters. First, the relationship between the electric flux density and the electric field intensity. For isotropic media, the linear relationship between the electric flux density D and the electric field intensity E is given by the Eq. 1.5a.

$$D = \varepsilon E \qquad (1.5a)$$

In the above equation, ε is the permittivity of the medium. Electric permittivity is the ability of a material to store electric potential energy in the presence of an external electric field. We will explore the permittivity more in detail later chapters.

Now, lets generalize the above system for a linear-time invariant system, and the equation becomes:

$$D(r, t) = \varepsilon(r, t) * E(r, t) \qquad (1.5b)$$

The $*$ operation indicates the convolution and r indicates any position vector. Once we get the Fourier transform of the above equation, it leads to the following equation where ω is the angular frequency.

$$D(r, \omega) = \varepsilon(r, \omega)E(r, \omega) \qquad (1.5c)$$

For anisotropic media, the permittivity $\varepsilon(r, \omega)$ becomes a tensor.

Similarly for magnetic fields, the linear relationship between the magnetic flux density and the magnetic field intensity is given by Eq. 1.6a.

$$B = \mu H \qquad (1.6a)$$

The parameter μ is the magnetic permeability of the medium. Magnetic permeability is the ability of a material to create an internal magnetic field when exposed to an external magnetic field. Like the electric fields, for a linear time-invariant system, this relationship becomes:

$$\boldsymbol{B}(\boldsymbol{r}, t) = \mu(\boldsymbol{r}, t){*}\boldsymbol{H}(\boldsymbol{r}, t) \tag{1.6b}$$

By getting the Fourier transform, the frequency domain relationship becomes:

$$\boldsymbol{B}(\boldsymbol{r}, \omega) = \mu(\boldsymbol{r}, \omega)\boldsymbol{H}(\boldsymbol{r}, \omega) \tag{1.6c}$$

Another important relationship is the one between the current density $\boldsymbol{J}$ and the electric field intensity $\boldsymbol{E}$. As shown in the Eq. 1.7a, these two parameters are related by the conductivity σ.

$$\boldsymbol{J} = \sigma\boldsymbol{E} \tag{1.7a}$$

Once we evaluate the Fourier transform of the linear time-invariant relationship of the current density and the electric field intensity, the frequency domain relationship becomes:

$$\boldsymbol{J}(\boldsymbol{r}, \omega) = \sigma(\boldsymbol{r}, \omega)\boldsymbol{E}(\boldsymbol{r}, \omega) \tag{1.7b}$$

The purpose of introducing the time and frequency domain relationships is to show that the parameters of electric and magnetic fields depend on the time, and frequency. And if the medium is anisotropic, the electric permittivity, magnetic permeability and the conductivity becomes tensors and depend on direction as well.

Throughout this book, we express the above frequency domain constitutive relationships as simple linear relationships. But its worth noting that when working with time-harmonic fields, the field parameters become functions of frequency as well.

Point Form of Maxwell's Equations

The above form of Maxwell's equations is known as the integral form. Using the Divergence theorem and Stokes's theorem we can derive their point form.

The divergence is defined as the net flux through an infinitesimal volume. In equation form it is defined as:

$$div\mathrm{D} = \lim_{\Delta V \to 0} \frac{\oint_s \boldsymbol{D}.\boldsymbol{ds}}{\Delta V} \tag{1.8}$$

Fig. 1.1 Divergence theorem: the vector summation of the flux through each infinitesimal element is equal to the flux through the closed surface shown in red

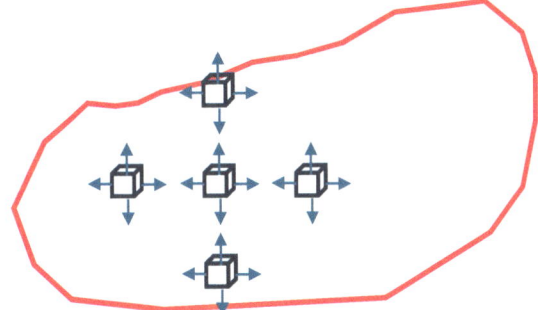

The divergence theorem says that the integration of a vector over a closed surface is equal to the integration of the divergence of that vector over a volume. Figure 1.1 shows the divergence theorem in graphical format.

Let's apply the divergence theorem to Gauss's Law for electric and magnetic fields. From the divergence theorem, for the vector $\boldsymbol{D}$,

$$\oint_s \boldsymbol{D}.ds = \int_v \nabla.\boldsymbol{D}.dv \tag{1.8a}$$

From the Gauss's Law for electric fields,

$$\oint_s \boldsymbol{D}.ds = Q_{enclosed} = \int_v \rho_v.dv \tag{1.8b}$$

In Eq. 1.8b ρ_v is the volume charge density or the charge per unit volume. Applying the divergence theorem in Gauss's Law,

$$\int_v \nabla.\boldsymbol{D}.dv = \int_v \rho_v.dv \tag{1.8c}$$

Eliminating the redundant volume integration will lead to the point form of Gauss's Law for electric fields.

$$\nabla.\boldsymbol{D} = \rho_v \tag{1.8d}$$

Similarly, we can obtain the point form of Gauss's Law for magnetic fields as:

$$\nabla.\boldsymbol{B} = 0 \tag{1.9}$$

Gauss's law for magnetic fields is also called the solenoidal law since the divergence of the magnetic flux density is zero.

Fig. 1.2 Stoke's theorem, the vector summation of all cross products over a closed surface is equal to the integration of a vector along a closed path shown in red

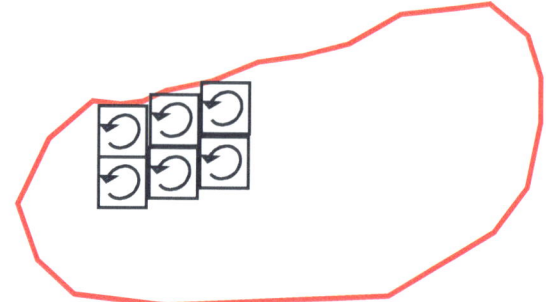

Let's get the point form of the Faraday's and Ampere's laws using the Stokes's theorem. Stoke's theorem says that the integration of a vector along a closed loop is equal to the integration of the curl of that vector field over a surface. Figure 1.2 shows the Stoke's theorem in graphical format.

Let's apply that to the electric field intensity.

$$\oint_l \boldsymbol{E}.d\boldsymbol{l} = \int_s \nabla \times \boldsymbol{E}.d\boldsymbol{s} \tag{1.10}$$

Now, let's apply Stokes's theorem to the Faraday's Law

$$\oint_l \boldsymbol{E}.d\boldsymbol{l} = \int_s \nabla \times \boldsymbol{E}.d\boldsymbol{s} = -\frac{\partial \phi_m}{\partial t} \tag{1.10a}$$

$$\int_s \nabla \times \boldsymbol{E}.d\boldsymbol{s} = -\frac{\partial \int_s \boldsymbol{B}.d\boldsymbol{s}}{\partial t} \tag{1.10b}$$

Assuming spatial and temporal independence we can write,

$$\int_s \nabla \times \boldsymbol{E}.d\boldsymbol{s} = -\int_s \frac{\partial \boldsymbol{B}}{\partial t}.d\boldsymbol{s} \tag{1.10c}$$

The point form of Faraday's law becomes:

$$\nabla \times \boldsymbol{E} = -\frac{\partial \boldsymbol{B}}{\partial t} \tag{1.10d}$$

Similarly, for the Ampere's Law,

$$\oint_l \boldsymbol{H}.d\boldsymbol{l} = \int_s \nabla \times \boldsymbol{H}.d\boldsymbol{s} \tag{1.11}$$

Applying, Stokes's theorem in Ampere's Law,

$$\int_s \nabla \times \boldsymbol{H}.d\boldsymbol{s} = \int_s \boldsymbol{J}.d\boldsymbol{s} + \int_s \frac{d\boldsymbol{D}}{dt}.d\boldsymbol{s} \qquad (1.11a)$$

And the point form of the Ampere's Law becomes:

$$\nabla \times \boldsymbol{H} = \boldsymbol{J} + \frac{d\boldsymbol{D}}{dt} \qquad (1.11b)$$

This book emphasizes how Maxwell's equations are applied to creating real-world applications. Some of these applications are widely used equations in physics and engineering to analyze electric and magnetic systems. The derivations and how the equations are applied are given with each application. Although there is an uncountable number of applications of Maxwell's equations, this book tried to capture a sample of those from various categories. Since the book is about applications the history of each application is omitted in some cases. The interested reader is suggested to refer to the references to learn the history of each application.

Reference

- Website: https://engineering.purdue.edu/wcchew/ece604f20/Lecture%20Notes/Lect7. pdf, date accessed: July 31, 2024.

Electrostatic Applications

2

Electrostatic applications use Gauss's Law for electric fields or Maxwell's first equation as their primary operating principle. This chapter shows some of the most common applications of Gauss's Law for electric fields and how to apply Gauss's Law in different coordinate systems.

Application 1: Capacitance of a Parallel Plate Capacitor

Figure 2.1 shows the configuration of a parallel plate capacitor. In a parallel plate capacitor, the top and bottom metal plates hold equal amounts of charges with opposite polarity. The medium between the capacitor plates is filled with a dielectric medium. Let's assume that the relative permittivity of the dielectric medium between the capacitor plates is ε_r, the voltage of the top plate is V_d, the potential of the bottom plate is zero, the surface area of a plate is S, and the surface charge density is ρ_s.

Let's find the capacitance of this parallel plate capacitor using Gauss's Law for electrostatic fields. To find the capacitance we need to find the total charge and the voltage between the plates. The basic equation for the capacitance is $C = \frac{Q}{V}$. First, let's enclose the top plate with a closed Gaussian surface and start with the Gauss's Law for electric fields.

$$\oint_s \boldsymbol{D}.\boldsymbol{ds} = Q_{enclosed} \tag{2.1}$$

© The Author(s), under exclusive license to Springer Nature Switzerland AG 2025
A. Maxworth, *One Hundred Applications of Maxwell's Equations*, Synthesis Lectures
on Electromagnetics, https://doi.org/10.1007/978-3-031-73784-8_2

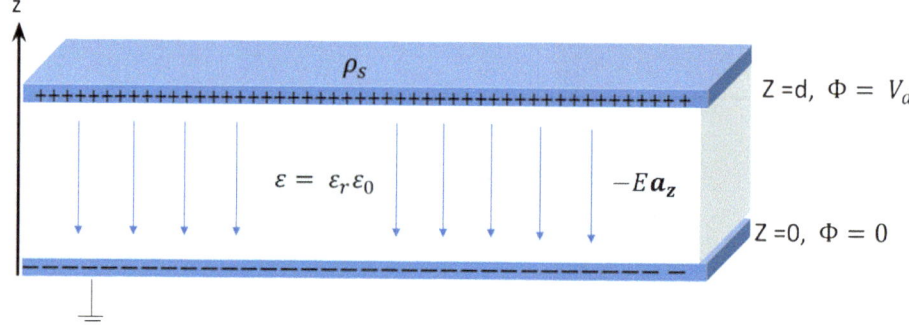

Fig. 2.1 The configuration of a parallel plate capacitor in Cartesian coordinates

In Eq. 2.1, D is the electric flux density and $Q_{enclosed}$ is the charge enclosed. When we look at the Gaussian surface, we can see that there is a net electric flux only through the bottom surface of the cuboidal Gaussian surface. The flux is along the $-a_z$ direction. Therefore, we need to consider the flux density in the $-a_z$ direction coming from the surface area facing the $-a_z$ direction. The total charge enclosed by the Gaussian surface is $\rho_s S$. Figure 2.2 shows the direction of the electric field intensity simulated by Ansys Maxwell 3D™ software for a parallel plate capacitor.

$$\int_s (-D_z a_z).(-ds a_z) = \rho_s S \tag{2.1a}$$

$$\int_s (-\varepsilon_r \varepsilon_o E_z a_z).(-ds a_z) = \rho_s S \tag{2.1b}$$

$$\varepsilon_r \varepsilon_o E_z S = \rho_s S \tag{2.1c}$$

$$E_z = \frac{\rho_s}{\varepsilon_r \varepsilon_o} \tag{2.1d}$$

$$E = -\frac{\rho_s}{\varepsilon_r \varepsilon_o} a_z \tag{2.1e}$$

Now, let's find the potential difference between the plates.

$$E = -\nabla \Phi \tag{2.1f}$$

In this capacitor configuration, the potential gradient exists only along the Z axis. Therefore,

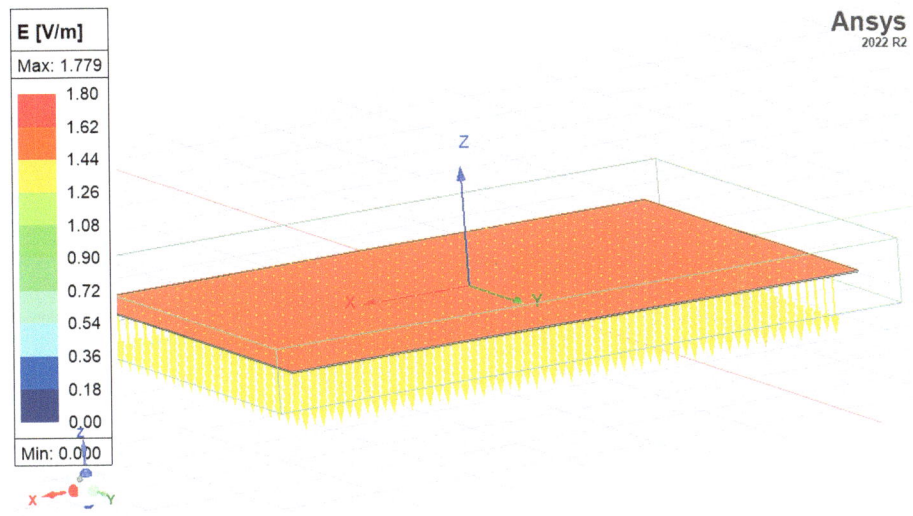

Fig. 2.2 Direction of the electric intensity of the parallel plate capacitor modeled using Ansys Maxwell 3D™ software

$$-\int_{z=d}^{0} E_z.dz = \Phi \qquad (2.1\text{g})$$

This gives the potential difference between the two plates as $\frac{\rho_s d}{\varepsilon_r \varepsilon_o}$.

Finally, plugging in the total charge and the potential difference between the two plates into the capacitance equation gives the capacitance as:

$$C = \frac{\varepsilon_r \varepsilon_o S}{d} \qquad (2.1\text{h})$$

Application 2: Capacitance of a Coaxial Cable

In this example, we will be applying Gauss's Law for static electric fields in the cylindrical coordinates. Coaxial cable is a derivative of the previous Trans-Atlantic Telegraph cable. The coaxial cables are used to transmit high-frequency radio signals. A coaxial cable has a metal core made from copper which conducts the signal, surrounded by a dielectric material enclosed by a copper mesh for electromagnetic shielding. The copper mesh is kept at ground potential. This entire configuration is enclosed by a plastic jacket for protection.

Fig. 2.3 Configuration of a coaxial cable in cylindrical coordinates

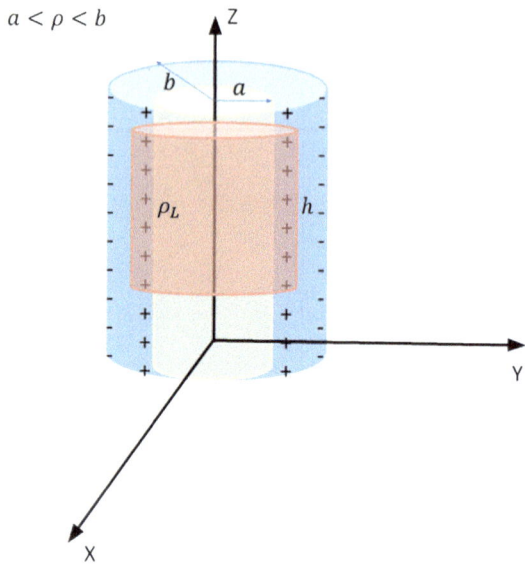

Figure 2.3 shows the configuration of a coaxial cable in cylindrical coordinates. Now, let's apply the Gauss's law for the coaxial cable. Let's consider the radius of the conductor is a and the radius of the copper mesh is b. The dielectric medium between the conductor and the copper mesh has a relative permittivity of ε_r.

In this case, we select a cylindrical Gaussian surface with a radius ρ. And height h.

$$\oint_s \boldsymbol{D}.\boldsymbol{ds} = Q_{enclosed} \tag{2.2}$$

In this configuration, the flux is coming through only from the $\boldsymbol{a}_\rho$ directed surface. The infinitesimal surface area element in the direction is $\rho \, d\phi \, dz$. Let's consider that the line charge density along the conductor is ρ_l. Hence the total charge enclosed by the Gaussian surface is $\rho_l h$. Therefore, the equation becomes:

$$\int_{z=0}^{h} \int_{\phi=0}^{2\pi} D_\rho \boldsymbol{a}_\rho.\rho \, d\phi \, dz \, \boldsymbol{a}_z = \rho_l h \tag{2.2a}$$

$$\int_{z=0}^{h} \int_{\phi=0}^{2\pi} \varepsilon_r \varepsilon_o E_\rho \boldsymbol{a}_\rho.\rho \, d\phi \, dz \, \boldsymbol{a}_z = \rho_l h \tag{2.2b}$$

$$E_\rho = \frac{\rho_l}{2\pi \varepsilon_r \varepsilon_o \rho} \tag{2.2c}$$

$$E = \frac{\rho_l}{2\pi \varepsilon_r \varepsilon_o \rho} \boldsymbol{a}_\rho \tag{2.2d}$$

Now, let's find the potential difference between the conductor and the copper mesh. Remember that the copper mesh is kept at ground potential.

$$E = -\nabla \Phi \tag{2.2e}$$

In this case, the potential gradient exists only in the $\boldsymbol{a}_\rho$ direction.

$$-\int_{\rho=b}^{a} E_\rho . d\rho = \Phi \tag{2.2f}$$

$$\frac{\rho_l}{2\pi \varepsilon_r \varepsilon_o} \{\ln(b) - \ln(a)\} = \Phi \tag{2.2g}$$

$$\frac{\rho_l}{2\pi \varepsilon_r \varepsilon_o} \ln\left(\frac{b}{a}\right) = \Phi \tag{2.2h}$$

Hence the capacitance of a coaxial cable is $\frac{2\pi \varepsilon_r \varepsilon_o h}{\ln\left(\frac{b}{a}\right)}$.

As shown in the equation above, the capacitance is a function of the length of the coaxial cable. Capacitance per unit length of a coaxial cable is:

$$C' = \frac{2\pi \varepsilon}{\ln\left(\frac{b}{a}\right)} \tag{2.2i}$$

Figure 2.4 shows the electric field generated by a coaxial cable in Ansys Maxwell 3D.

Application 3: Electrostatic Precipitator

Electrostatic precipitator uses the principle of Gauss's Law for electrostatic fields to extract harmful particles from exhaust air. The working principle is as follows:

Step 1: The exhaust air including dust and other molecules enters the precipitator.

Step 2: the exhausted air passes through a mesh at a very high negative potential. As the exhaust air passes through this highly negative grid, the dust particles pick up a negative charge.

Step 3: the exhaust air passes through highly positively charged metallic plates. As the air passes through these metallic plates the negatively charged dust particles are attracted towards the highly positive metallic plates and become neutralized. Those attached particles are brushed off later and the clean exhaust air exists in the precipitator.

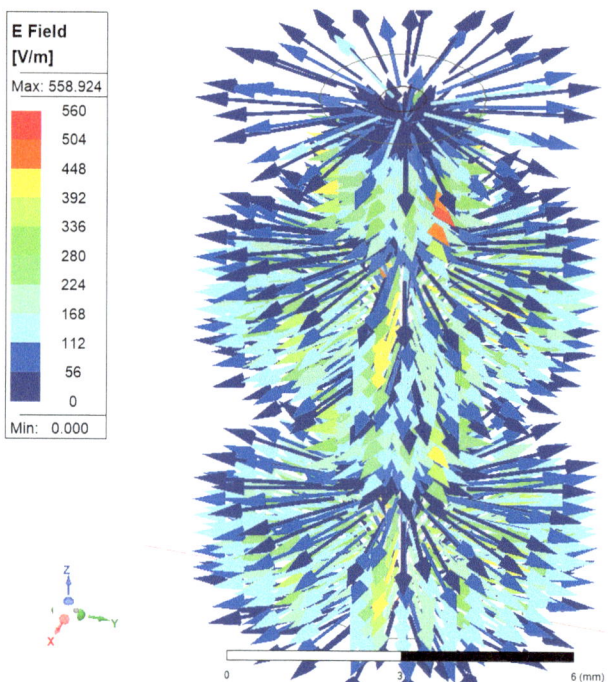

Fig. 2.4 Direction of the electric field intensity of a coaxial cable modeled in Ansys Maxwell 3D

Figure 2.5 shows an electrostatic precipitator. Gauss's law is applied in designing electrostatic precipitators to determine the potential of the metallic grid, the potential of the collective plates, and the surface area.

Application 4: Electro-Filter or the Electrostatic Drum Separator

The electro-filter shown in Fig. 2.6 uses a similar concept as the electrostatic precipitator. The purpose of the electro-filter is to separate the charged and uncharged particles. In the process first, a mix of negatively charged and uncharged particles are released onto the surface of the drum. At this entry stage, a very high positive voltage is induced onto the particles through a corona discharge. This corona discharge neutralizes the previously negatively charged particles while positively charging the particles that did not initially carry a charge. As the drum spins around its axis, the neutral particles fall due to gravity and are collected at the bottom of the drum while the positively charged particles are brushed from the surface of the drum and collected separately.

Fig. 2.5 Electrostatic
precipitator

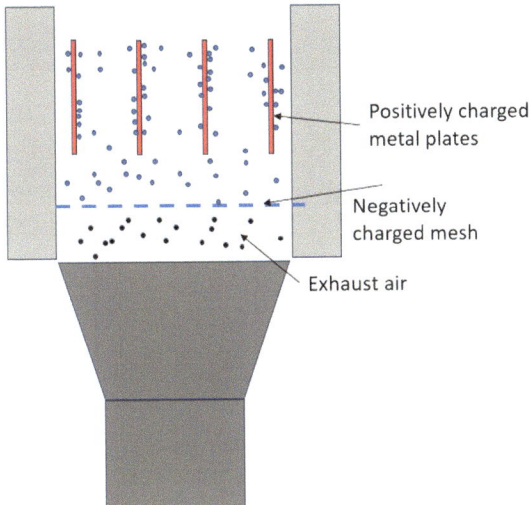

Positively charged
metal plates

Negatively
charged mesh

Exhaust air

Fig. 2.6 The electro-filter or
the drum separator

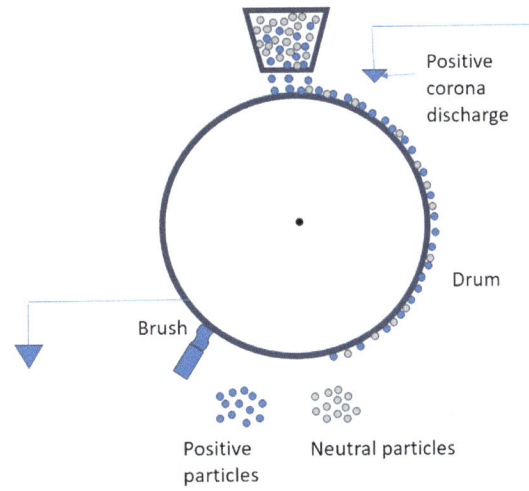

Positive
corona
discharge

Drum

Brush

Positive
particles

Neutral particles

Application 5: Xerographic Printing

Xerographic comes from the Greek words dry-writing. In Xerographic printing, a dry power is used as a toner. The printer is comprised of an aluminum drum coated with selenium. Selenium is a photoconductor that conducts when exposed to light. The xerographic printing can be explained in 5 steps.

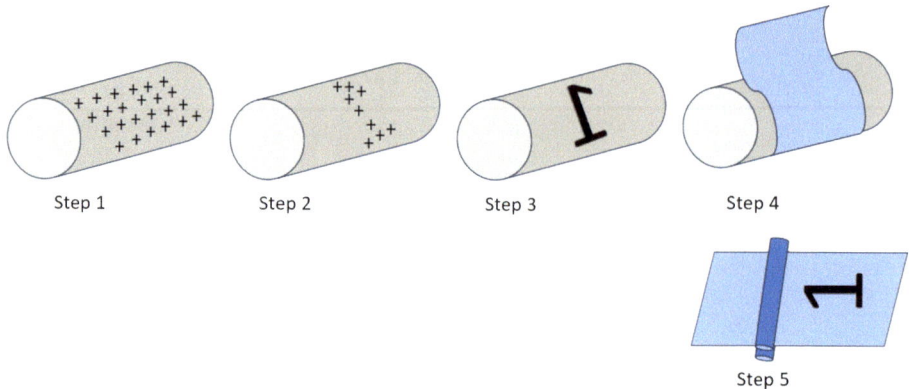

Step 1   Step 2   Step 3   Step 4

Step 5

Fig. 2.7 The steps in xerographic printing

Figure 2.7 shows the five steps of xerographic printing.

Step 1—charging: the selenium coating on the aluminum drum is given a positive charge.

Step 2—exposure: a mirror image of the image to be printed is imposed onto the drum using laser beams and mirrors. At this point, the areas of selenium that are in the dark retain their positive charge, and the areas exposed to light become conductive hence losing their charge.

Step 3 development: as the drum rolls around its axis, the negatively charged toner is attached to the positively charged areas of the drum.

Step 4—Transfer: as the drum rolls over the paper, the toner is transferred onto the paper.

Step 5—fusing: using heat and pressure rolls the print is infused onto the paper.

Application 6: Electrostatic Coating

Electrostatic coating uses the principle of Gauss's Law. In electrostatic coating, the negatively charged paint particles are sprayed on the positively charged surfaces. This coating technique is widely used when spray painting metallic surfaces. Given that Gauss's law states that charge is proportional to flux, hence one charge produces one flux. Hence, one positive charge will attract one negatively charged paint particle. The electrostatic coating is preferred over regular hand-painting given that this method produces a more even coating.

Application 7: Van Der Graff Generator

The Van der Graff generator (Fig. 2.8) was originally developed by the physicist Robert J Van der Graff to supply power to particle accelerators. Currently, it is used as a classroom demonstration tool for corona discharge. Let's study the operation of the Van der Graff generator step by step starting from the bottom roller.

The bottom roller of the Van der Graff generator is made with a negative material on the triboelectric series such as silicon. When this roller rubs against rubber which is neutral in the triboelectric series, the electrons of the atoms are transferred to the silicon roller while the positive ions are transferred to the rubber belt.

This highly negatively charged silicon roller starts breaking down the neutral atoms in the air surrounding it. Due to its high negativity, the positive ions break from their electrons and are attracted to the silicon roller. However, these positive ions cannot get attached to the silicon roller due to the rubber belt. And these positive ions are being carried away by the rotating belt. At the same time, the electrons are repelled away and these electrons land on the metallic brush at the bottom.

The electrons on the metallic brush at the bottom get repelled away from the highly negatively charged silicon roller. Therefore, there is a relatively positive charge at the tips of the bottom brush. The electrons stripped away from the air molecules land on the tips of this bottom brush which again being repelled away towards the grounding rod.

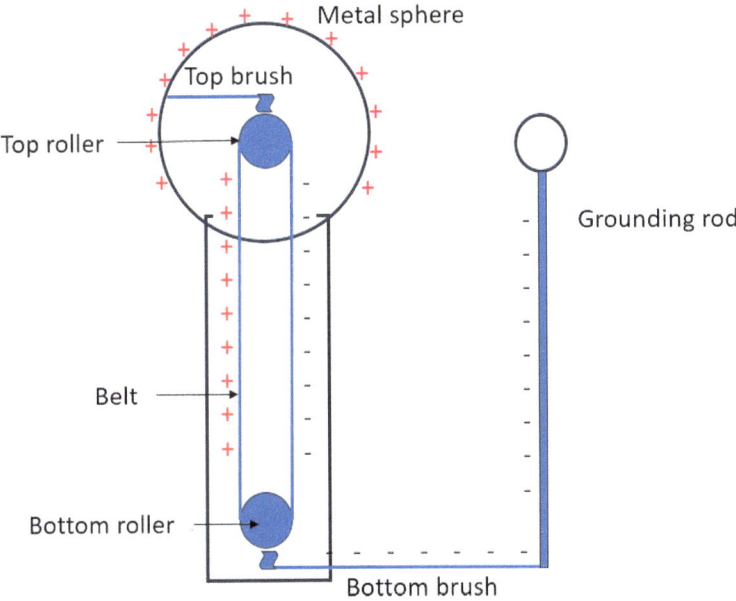

Fig. 2.8 The Van der Graff generator

Now, let's consider the upper roller which is made of a positive material of the tri-boelectric series such as nylon. In this case, the upper roller is charged more positively compared to the rubber belt. As the rubber belt brings more positive charges toward the upper roller, those positive charges are repelled away from the highly positive upper roller. At the same time, the electrons on the tips of the upper metallic brush are moved towards the brush. The positive charges on the brush are then attracted towards the upper metallic brush which is connected to the metallic sphere. Due to the strong repelling force acting on the positive charges from the upper roller, these positive charges are pushed towards the metallic sphere. At the same time the air molecules surrounding the upper roller and the upper brush break down landing the positive ions on the brush and the electrons or the negative ions on the belt.

When the grounding rod is brought close to the metallic sphere, the potential difference creates a strong electric field similar to a mini lightning strike.

References

- Website: https://www2.physics.ox.ac.uk/accelerate/resources/demonstrations/van-de-graaff-generator, accessed: June 25, 2024.
- Website: https://science.howstuffworks.com/transport/engines-equipment/vdg4.htm, accessed: June 25, 2024.

Application 8: Electrostatic Shielding

Electrostatic shielding protects, instruments, components, and areas from static electric charges. This is done by covering or laminating the outer surface of the object that needs to be shielded. The electrostatic shielding process can be explained using the following three figures created on Ansys Maxwell 3D. Figure 2.9a shows the electric field intensity inside an air cavity that needs to be shielded from electrostatic fields. It is covered by a shell made with a dielectric material polyvinyl chloride (PVC). The PVC shell contains a charge of 1C distributed evenly across its volume. The net charge inside the air cavity is zero (marked in dark blue) from Gauss's law since the fields from the charges located opposite to each other cancel out.

Figure 2.9b shows the same air cavity plus the PVC shell configuration with another charge of 2C in its proximity. In this case, the electric field intensity within the air cavity is non-zero due to the superposition of electric fields: i.e. every charge in space creates an electric field independent of other charges present. In this scenario, the air cavity is not electrostatically shielded by the PVC covering.

Figure 2.9c shows the same configuration as Fig. 2.9b, but instead of the PVC covering, the air cavity is covered with a copper layer. In this case, the electric field inside the air cavity is zero again, making it electrostatically shielded. Given that copper is a good

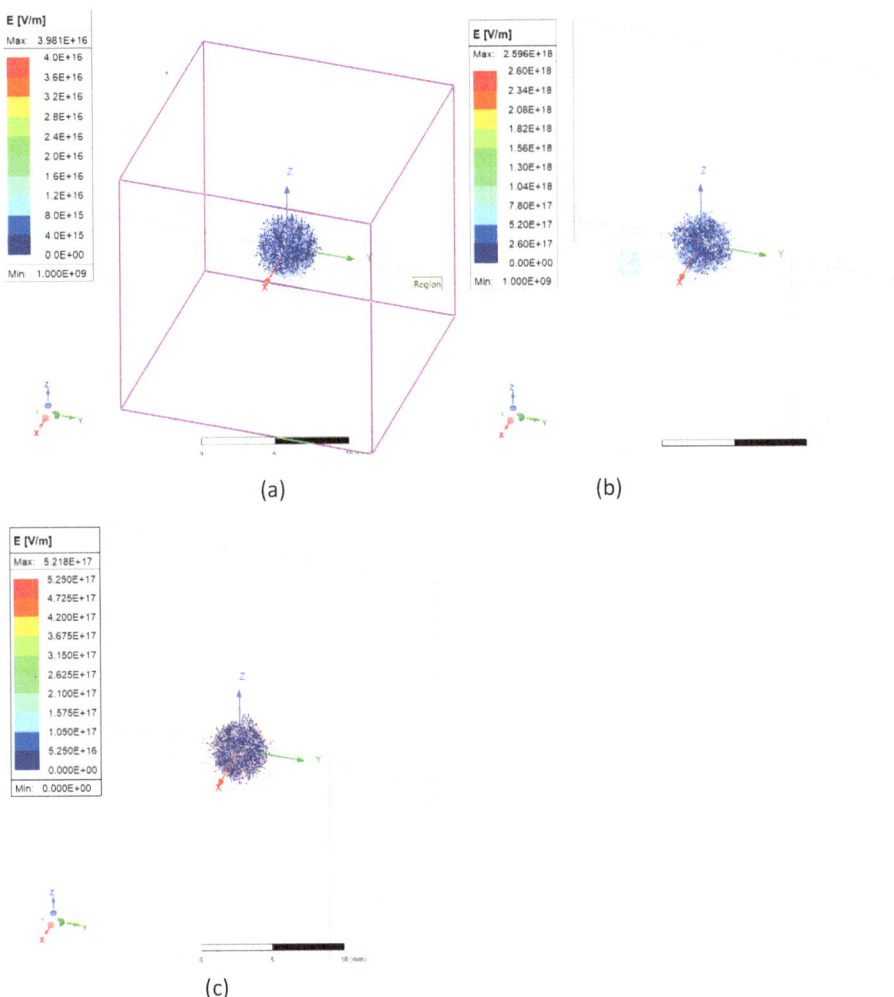

Fig. 2.9 a A PVC shell with a surface charge of 1C. **b** An external charge of 2C is brought near the charged PVC shell. **c** Instead of the PVC shell a copper shell is used to shield the external charge

conductor, in the presence of an electric field, it mobilizes its electrons to the surface to shield the external electric field. In this case, the external charge is positive, hence the electrons gather on the surface of the copper layer facing the external charge. As a result of electrons moving the copper layer on the opposite side of the external charge contains a positive potential.

Electrostatic shielding is used vastly in printed circuit board manufacturing and the electronics industry.

Application 9: Faraday Cage

Faraday Cage is named after the British experimentalist Michael Faraday. The Faraday cage is a great example of electromagnetic shielding. The Faraday cages may vary in size depending on the application. Faraday cages can be used to shield the interior of the cage from outside such as testing chambers or to avoid electromagnetic leakage from the interior to the exterior of the cage, for example: microwave oven.

The exterior of a Faraday cage is made with conductive materials such as copper, silver, or aluminum for electric shielding. At equilibrium and when no external fields are present, there are equal numbers of electrons and protons maintaining a uniform potential within all points of the conductor. When the Faraday cage is exposed to an external electric field the electrons attract to the positive electric field shielding the field from entering the cage. As a result of electrons flowing to one side of the cage, the opposite side of the cage has an excess of positive charges creating a potential difference. As shown in the Fig. 2.10.

For magnetic shielding, the Faraday cages are laminated with high-permeable material such as soft ferrite. High-permeable material can induce internal magnetic fields when exposed to an external field—warping the external magnetic field. Hence the interior of the Faraday cage is shielded from the magnetic field as well.

Faraday cages can be made either with metallic sheets or metallic nets. In both cases, it is important to avoid having openings greater than 1/10 of the smallest wavelength of the electromagnetic fields that need to be shielded.

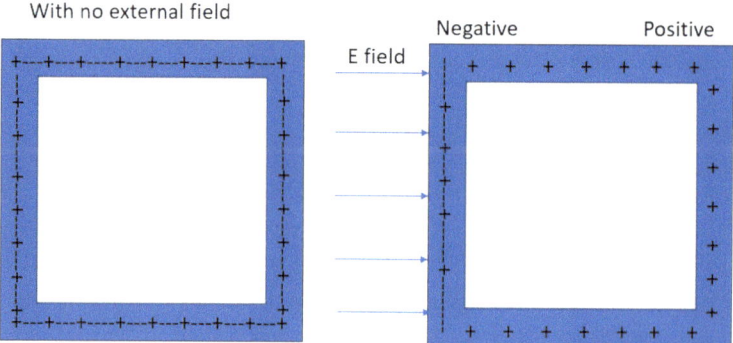

Fig. 2.10 Electrostatic charge distribution on a Faraday cage with and without an external electric field

References

- https://www.gamry.com/application-notes/instrumentation/faraday-cage/#:~:text=Gen eral%20rule%20of%20thumb%20is,be%20smaller%20than%203%20mm.   Accessed: June 25, 2024.
- https://backyardbrains.com/experiments/faraday#:~:text=A%20Faraday%20cage% 20is%20a,interfering%20with%20our%20neural%20recordings accessed: June 25, 2024.

Application 10: Electrostatic Energy Stored in a Capacitor

Electrostatic work indicates the work needed in moving a group of charges a certain distance. The average electric work is defined as:

$$W_{electric} = \frac{1}{2} \int_l F_e . ldl \tag{2.3}$$

The division by 2 is because when charge Q_1 is arranged against charge Q_2, no work is necessary to arrange charge Q_2 against charge Q_1.

The electric field intensity is the force acting on a unit charge and the total charge. Hence, $F_e = QE$. And from the Gauss's Law, $Q = \oint_s D.ds$. Hence,

$$W_{electric} = \frac{1}{2} \oint_s \int_l E.Ddlds = \frac{1}{2} \int_v E.Ddv \tag{2.3a}$$

In the parallel plate capacitor example, we found that the electric field intensity within the dielectric is:

$$E = -\frac{\rho_s}{\varepsilon_r \varepsilon_o} a_z \tag{2.3b}$$

Let's use the equation for the stored electric energy to find the electrostatic energy stored in a capacitor.

$$W_{electric} = \frac{1}{2} \int_v E.Ddv = \frac{1}{2} \int_v \varepsilon |E|^2 dv \tag{2.3c}$$

$$W_{electric} = \frac{1}{2} (\int_v \varepsilon_r \varepsilon_o \left| -\frac{\rho_s}{\varepsilon_r \varepsilon_o} \right|^2 dv) \tag{2.3d}$$

$$W_{electric} = \frac{1}{2}\varepsilon_r\varepsilon_o\left(\frac{\rho_s}{\varepsilon_r\varepsilon_o}\right)^2 Sd \qquad (2.3e)$$

Now, let's rearrange the terms such that we can get the above equation in the capacitance and potential difference.

$$W_{electric} = \frac{1}{2}\frac{\varepsilon_r\varepsilon_o S}{d}\left(\frac{\rho_s d}{\varepsilon_r\varepsilon_o}\right)^2 \qquad (2.3f)$$

$$W_{electric} = \frac{1}{2}C\Phi^2 \qquad (2.3g)$$

Reference

- Website: https://physics.stackexchange.com/questions/497622/why-is-there-a-1-2-in-the-expression-for-electrostatic-energy-u, accessed June 25, 2024.

Application 11: Derivation of the Coulomb's Law

Coulomb's law helps determine the electric field intensity and electric scalar potential given an arbitrary charge distribution. This law has its applications in analyzing antenna structures including linear wire antennas, reflector antennas and aperture antennas. The Gauss's law in point form for electric fields states:

$$\nabla.\mathbf{D} = \rho_v \qquad (2.4)$$

If the charge distribution is in air, we can write the electric field equation as:

$$\nabla.\mathbf{E} = \frac{\rho_v}{\varepsilon_o} \qquad (2.4a)$$

$$\mathbf{E} = -\nabla\Phi \qquad (2.4b)$$

Also, let's assume that the volume charge distribution is a function of location.

$$\nabla^2\Phi = -\frac{\rho_v(x',y',z')}{\varepsilon_o} \qquad (2.4c)$$

At this point, let's use the Green's theorem. Green's theorem says that Green's function is the response when a linear operator is applied to a Dirac-delta function. In the above equations, the Laplacian operator is the linear operator, the vector magnetic potential is the response, and the current densities are the sources.

$$\mathbf{r} = x\mathbf{a}_x + y\mathbf{a}_y + z\mathbf{a}_z \qquad (2.4d)$$

$$r' = x'_x^a + y'_y^a + z'_z^a \tag{2.4e}$$

Equations 2.4d and 2.4e represent the position vectors of the observer and the source. Let

$$R = r - r' = (x - x')a_x + (y - y')a_y + (z - z')a_x \tag{2.4f}$$

$$R = |r - r'| = \sqrt{(x - x')^2 + (y - y')^2 + (z - z')^2} \tag{2.4g}$$

The Green's function for the Laplacian operator is the well-known:

$$G(r, r') = -\frac{1}{4\pi |r - r'|} \tag{2.4h}$$

In the above Green's function, r is the observer location, and r' is the source location. The convolution between the Green's function and the source functions should provide the responses required. Hence:

$$\Phi = \frac{1}{4\pi \varepsilon_o} \int_v \frac{\rho_v(x', y', z')}{|r - r'|} dv \tag{2.4i}$$

The equation for the electric field intensity becomes:

$$E = -\frac{1}{4\pi \varepsilon_o} \int_v \rho_v(x', y', z') \nabla \cdot \frac{1}{|r - r'|} dv \tag{2.4j}$$

$$\frac{\partial}{\partial x} \left(\frac{1}{|r - r'|} \right) = -\frac{1}{2} \frac{2(x - x')}{\left(\sqrt{(x - x')^2 + (y - y')^2 + (z - z')^2} \right)^3} \tag{2.4k}$$

$$\frac{\partial}{\partial x} \left(\frac{1}{|r - r'|} \right) = -\frac{(x - x')}{|r - r'|^3} \tag{2.4l}$$

$$\frac{\partial}{\partial y} \left(\frac{1}{|r - r'|} \right) = -\frac{(y - y')}{|r - r'|^3} \tag{2.4m}$$

$$\frac{\partial}{\partial z} \left(\frac{1}{|r - r'|} \right) = -\frac{(z - z')}{|r - r'|^3} \tag{2.4n}$$

$$E = \frac{1}{4\pi \varepsilon_o} \int_v \rho_v(x', y', z') \frac{R}{|R|^3} dv \tag{2.4o}$$

If the source is a surface charge distribution, the above equation modifies to:

$$E = \frac{1}{4\pi\varepsilon_o} \int_s \rho_s(x', y', z') \frac{R}{|R|^3} ds \tag{2.4p}$$

And for a line charge distribution:

$$E = \frac{1}{4\pi\varepsilon_o} \int_l \rho_l(x', y', z') \frac{R}{|R|^3} dl \tag{2.4q}$$

Reference

- Website: https://www.youtube.com/watch?v=lXF7HDVrCRI, accessed June 25, 2024.

Application 12: Electric Field from an Electric Dipole

Let's consider the two equations for the electrostatic potential and the electric field intensity derived above. The electric dipole analysis was fundamental in antenna analysis including the infinitesimal dipole, Hertzian dipole and short dipole.

$$\Phi = \frac{1}{4\pi\varepsilon_o} \int_v \frac{\rho_v(x', y', z')}{|r - r'|} dv \tag{2.5a}$$

$$E = \frac{1}{4\pi\varepsilon_o} \int_v \rho_v(x', y', z') \frac{R}{|R|^3} dv \tag{2.5b}$$

For point charges the two above equations reduce to the following with Q being the point charge.

$$\Phi = \frac{Q}{4\pi\varepsilon_o R} \tag{2.5c}$$

$$E = \frac{Q}{4\pi\varepsilon_o R^2} a_R \tag{2.5d}$$

a_R is the unit vector in the direction of R. And $a_R = \frac{R}{|R|}$. The above two equations can be derived using the Gauss's law as well.

Let's apply the above equations to get the potential difference due to a dipole. In this case, there are two charges one with a charge $+q$ and $-q$ separated by a distance d as shown in the Fig. 2.11.

The potential at point P due to the two charges is:

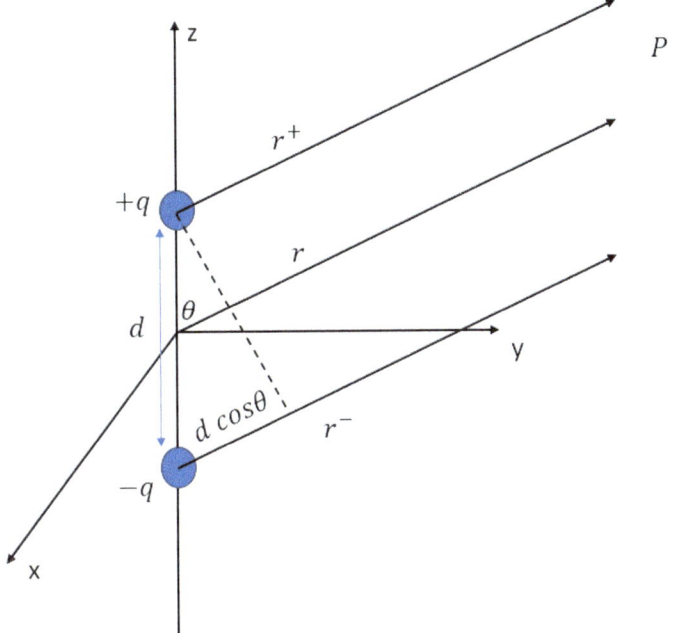

Fig. 2.11 The electric dipole

$$\Phi = \frac{q}{4\pi\varepsilon_o r^+} - \frac{q}{4\pi\varepsilon_o r^-} = \frac{q}{4\pi\varepsilon_o} \frac{\left(r^- - r^+\right)}{r^+ r^-} \qquad (2.5\text{e})$$

In the far field, the $r^+ \approx r^- \approx r$, and $r^- - r^+ = d\cos\theta$

$$\Phi = \frac{qd\cos\theta}{4\pi\varepsilon_o r^2} \qquad (2.5\text{f})$$

$$E = -\nabla\Phi \qquad (2.5\text{g})$$

In spherical coordinates, the gradient operator is:

$$\nabla = \frac{\partial}{\partial r}a_r + \frac{1}{r}\frac{\partial}{\partial\theta}a_\theta + \frac{1}{r\sin\theta}\frac{\partial}{\partial\phi}a_\phi \qquad (2.5\text{h})$$

The electric field intensity becomes:

$$E = \frac{qd\cos\theta}{2\pi\varepsilon_o r^3}a_r + \frac{qd\sin\theta}{4\pi\varepsilon_o r^3}a_\theta \qquad (2.5\text{i})$$

Application 13: Faraday Cup Electrometer

A Faraday cup is a metallic cup (like a Faraday cage) used to measure the charge in air or the flow of a chemical. The 2.12 below shows the construction of the Faraday cup. The current total charge is measured by the amount of current flow. The operation of the Faraday cup can be explained by the simple equation below:

$$Nq = It \tag{2.6}$$

Since the current is the rate of change of charge, when we observe the current for a certain period, it gives the total charge. By dividing the total charge by the charge of a proton, the total number of protons present within the flow can be found.

$$\frac{N}{t} = \frac{I}{q} \tag{2.6a}$$

In the above equation, N is the total number of protons, I is the current, t is the observational period and q is the charge of a proton (Fig. 2.12).

As the airflow enters the cup, the electrons are attracted to the electron suppression plates. The grounded shield prohibit the ions entering the cup from outside other than the air inlet. Once the airflow is inside the cup, the ions are transferred to the metallic Faraday cup, and this induces a current that is measured by the electrometer.

References

- Website: https://www.thermofisher.com/us/en/home/industrial/spectroscopy-elemental-isotope-analysis/spectroscopy-elemental-isotope-analysis-learning-center/trace-elemental-analysis-tea-information/inductively-coupled-plasma-mass-spectrometry-icp-ms-information/icp-ms-systems-technologies.html accessed June 25, 2024.
- Website: https://faradaym.weebly.com/faradays-cup.html accessed June 25, 2024.

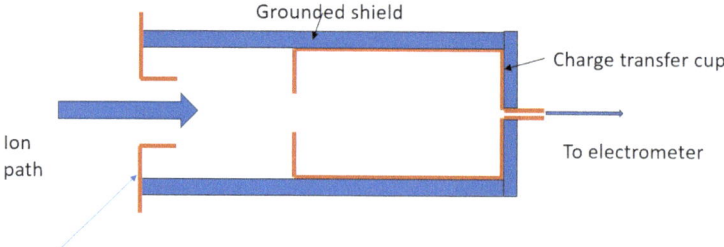

Fig. 2.12 The Faraday cup electrometer

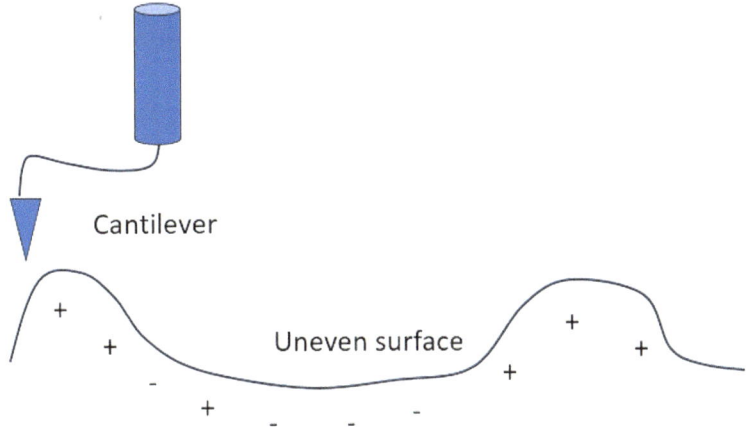

Fig. 2.13 The electrostatic force microscopy cantilever position

Application 14: Electrostatic Force Microscopy

Electrostatic force microscopy (EFM, shown in Fig. 2.13) is a method of analyzing the surface potential at the *nm* scale. This technique falls under non-contact atomic force microscopy (AFM). Similar to all AFM techniques this method also uses a cantilever. But rather than having the tip touch the sample, here the tip is kept at a constant distance from the sample.

This is a two-pass technique; hence each sample will be scanned twice. In the first scan, the Van der Waal forces are measured. Van der Waal forces are the chemical forces. Hence during the first scan, the goal is to find the height of constant Van der Waal forces and this line of constant Van der Waal forces gives the topography of the sample.

The second scan is to find the surface charge of the sample. Finding these surface charges is important in the electronic semiconductor manufacturing industry. In this case, the tip is at a DC bias voltage (V_{DC}) with a superimposed AC signal (V_{ac}). If the sample voltage is V_s, the voltage between the tip and the sample would be:

$$v(t) = V_{DC} + V_{ac}\sin(\omega t) - V_s \tag{2.7a}$$

The electrostatic force can be calculated from the electrostatic portion of the Lorentz force:

$$F_{electrostatic} = qE \tag{2.7b}$$

Now, let's consider the relationships $q = Cv(t)$ and $E = v(t)/d$

$$F_{electrostatic} = Cv(t).\frac{v(t)}{d} = \frac{Cv^2(t)}{d} \tag{2.7c}$$

In the above equation, C is the capacitance between the tip and the sample, and d is the distance between the tip and the sample. This equation gives:

$$F_{electrostatic} = \frac{C}{d}\left[(V_{DC} - V_S)^2 + \frac{1}{2}V_{ac}^2\right] + 2\frac{C}{d}(V_{DC} - V_S)V_{ac}\sin(\omega t)$$
$$- \frac{1}{2}\frac{C}{d}V_{ac}^2\cos(2\omega t) \tag{2.7d}$$

The first portion of this signal is known as the DC cantilever deflection signal and that can be read directly using either a laser position sensor or another method. The AC portions of the signals are read by sending the signal to a lock-in amplifier. By analyzing the three portions of the above signal the sample surface voltage can be obtained.

References

- https://www.parksystems.cn/park-spm-modes/93-dielectric-piezoelectric/228-electric-force-microscopy-efm accessed June 25, 2024.
- https://pubs.acs.org/doi/10.1021/nn5041476 accessed June 25, 2024.

Practice Problems

1. Consider the Van der Graff generator shown in Fig. 2.14.
   a. The outer surface of the bigger sphere contains a constant positive surface charge of 0.3 C/m^2 (constant surface charge density). Calculate the total charge on that surface.
   b. Using a suitable law determine the electric field intensity at point P due to the surface charge on the bigger sphere.
   c. The outer surface of the smaller sphere needs to be given a negative charge. Assuming the surface charge density on the smaller sphere as $-\rho_s$ C/m^2, expresses the electric field intensity at point P, only due to the charge distribution on the surface of the small sphere.
   d. Using the expressions obtained in parts b and c, calculate the value of ρs, such that the electric field intensity at point P becomes zero.
2. The rectangular version of the coaxial cable is called a strip line (shown in Fig. 2.15). Here the inner conductor is given a total static charge of $+Q$ and the outer conductor is given a total charge of $-Q$. The direction of the electric field is marked in the diagram. The space between the two conductors is filled with a material with a dielectric constant of 2.54.
   a. By selecting a suitable Gaussian surface, calculate the electric field intensity between the conductors.
   b. Hence calculate the electrostatic potential difference between the conductors.
   c. How much of a capacitance can be observed between the conductors?

Fig. 2.14 The Van der Graff generator for the problem 1

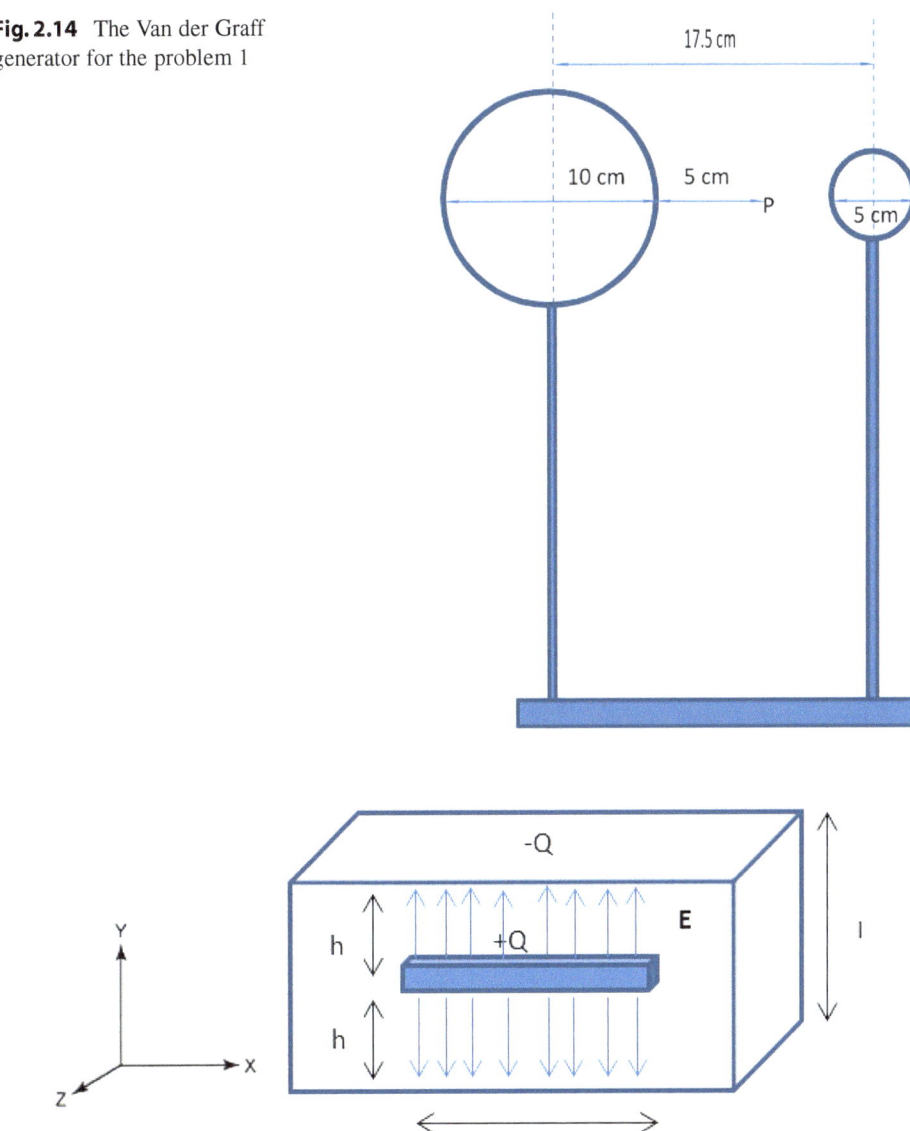

Fig. 2.15 The strip line

3. Lightning occurs due to the potential difference between a cloud and the ground. Figure 2.16 shows a simplified system.
   a. Assuming the charge cloud is spherical with a volume charge density of -0.03 C/m3, calculate the total charge stored within a spherical cloud with a radius of 1 m.

b. Using a suitable Gaussian surface, derive an expression for the radial electric field intensity (Er) in the region between the cloud and the ground.

Hint: Here you are giving an expression for the electric field intensity at radially r (1 m< r <100 m) distance away. You may assume that the cloud is centered at the origin and the medium is air.

c. Using your answers for parts (a) and (b), calculate the potential difference between the cloud and the ground.

Hint: $\int \frac{1}{r^2} dr = -\frac{1}{r}$ and the limits of integration would be from 1 to 100 m.

d. Hence calculate the capacitance between the charge cloud and ground.

e. Calculate the amount of electrostatic energy stored between the spherical cloud and the ground. Here the volume of interest would be the volume between the two spheres.

4. Electrostatic Precipitator (ESP) is a device used to remove particle waste from exhaust air. The electrostatic precipitator shown Fig. 2.17 contains two hollow cylinders. The diameter of the inner cylinder is 30 cm and the diameter of the outer cylinder is 1 m. In both cylinders the wall thicknesses are negligible. The height of the device is 1 m. The outer surface of the inner cylinder is given an evenly distributed surface charge density of $\rho_s = -0.3$ C/m^2.

You may assume that both cylinders are axially centered on the z-axis and that the medium inside the cylinders is air. Since both cylinders are hollow you may ignore the top and bottom surfaces.

Exhaust air with waste particles is blown from below through the cylinders. The waste particles (typically coal ash) are positively charged. The positively charged ash particles are attached to the inner cylindrical surface which is negatively charged. Hence those are removed from the exhaust air. Note that the figure shows the direction of the airflow, but the electric field exists in the radial direction.

a. Consider the situation just before the exhaust air is blown through the cylinders. If the outer surface of the inner cylinder is given an evenly distributed charge density of $\rho s = -0.3$C/m^2, calculate the total amount of charges distributed on the surface of the inner cylinder.

Fig. 2.16 The simplified diagram for cloud to ground lightning

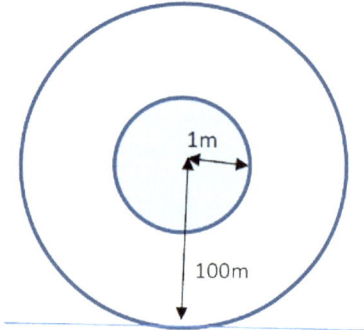

1m

100m

Fig. 2.17 An electrostatic precipitator with two concentric cylinders

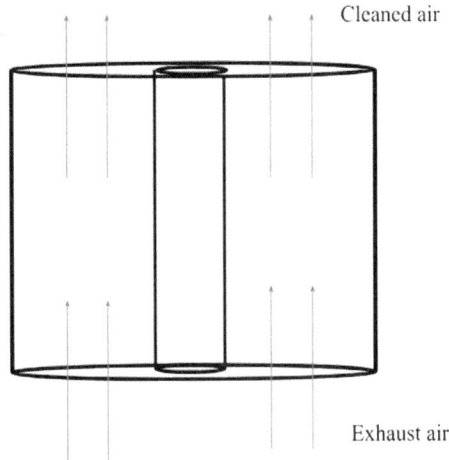

b. Before exhaust air is blown through the two cylinders, calculate the radial electric field intensity (E) at a radial distance of 20 cm from the surface of the inner cylinder due to the negative charge accumulation given in part a.

c. Now exhaust air is blown through the cylinders and here positively charged ash molecules get attached to the negative charge on the inner cylinder and become neutral. If each ash molecule has a charge of 2×10^{-5} C and 5000 of those were attached to the inner cylindrical surface how much is the net charge on the surface of the inner cylindrical surface? You may assume that the ash particles are evenly attached to the surface.

d. What can you say about the magnitude of the radial electric field intensity (E) 20 cm radially away from the surface of the inner cylinder once the ash particles were attached to the surface of the inner cylinder, will it be higher or lower compared to the answer you obtained in part b.

5. Electrostatic Coating.

Electrostatic coating is an efficient coating method for metal surfaces. In this method, the metal surface to be coated is connected to the ground and the negatively charged paint molecules are projected towards the metal surface. The negatively charged paint molecules are attracted to the metal surface due to electrostatic forces. Once in contact with the metal surface, the negatively charged paint molecules get evenly distributed among the surface due to the repelling forces from the adjacent paint molecules.

a. A cylindrical paint container with a diameter of 30 cm and a height of 30 cm contains 2×1016 paint molecules. And each paint molecule has a $-2e$ charge (e is the charge of an electron). What is the total charge inside the paint container in Coulombs?

b. Assuming that the paint molecules are densely and evenly distributed inside the container, calculate the volume charge density inside the paint container, for which the dimensions are given in part a.

c. Selecting an appropriate Gaussian surface calculate the radial electric field intensity on the inner cylindrical surface of the paint container. You may assume that the thickness of the paint container is negligible, the relative permittivity of paint is 5 ignoring the flux from the top and bottom.

d. Given the electric field intensity inside the paint container is -500 MV/m, how much polarization can be observed inside the paint container? The relative permittivity of paint is 5.

e. The entire volume of paint in the container was used to coat a spherical metal surface with a radius of 50 cm. What is the surface charge density on the coated metal surface?

6. Electrostatic insulation.

Mercury is the only conductive metal in liquid form at room temperature. One form of mercury is mercuric (Hg^{2+}) which has two protons in its valance band making it an ion. The height of the beaker carrying mercury is 15 cm, the diameter is 8 cm, and the mercury is filled up to a height of 3 cm.

a. If there are approximately 1024 atoms in one cubic centimeter, calculate the number of mercuric atoms inside the beaker.

b. Based on your answer in part a, calculate the total charge inside the beaker in coulombs.

c. A student built a semi-spherical (half a sphere) plastic enclosure with a radius of 15 cm to cover chemicals. Determine the electric flux density on the outer surface of the enclosure.

d. Based on your answer in part c, calculate the electric field intensity on the outer surface of the enclosure.

e. If the enclosure was built with copper (a very good conductor) how much should be the electric field intensity on the outer surface of the enclosure?
Justify your answer. Part e does not require any calculations.

7. Electro-filter or the electrostatic drum separator

a. Initially, the hopper contains 1 million (106) particles including 0.9 million neutrons and 0.1 million electrons. How much is the initial total charge inside the hopper in Coulombs?

b. Once the corona discharge neutralizes the electrons and imposes a positive charge on the neutrons (in other words, all the neutrons will be converted to protons) how much will be the net charge on the drum in Coulombs?

c. Assume that the net charge you calculated in part b, is evenly distributed on the full curved surface of a cylindrical drum with radius 10 cm and depth (height) 1 m. How much is the surface charge density on the cylindrical drum?

Make sure to rotate the system such that the depth is along the z-axis, and the radius of the cylinder is along the radial direction in a cylindrical coordinate system.

d. Using a suitable Gaussian surface calculate the electric flux density D, 1 m radially away from the center of the cylindrical drum. The only flux is through the curved cylindrical surface of the drum.

e. If the entire system is in air, how much is the electric field intensity (E) 1 m radially away from the center of the cylindrical drum?

8. Capacitance inside a diode.

Diodes are important semiconductor devices with a large number of applications. Diodes are constructed by attaching a p-type and an n-type semiconductors. For this question, you do not need to know the semiconductor theory. Here, we are interested ONLY in the depletion region of a diode or the region at the junction, since it acts as a capacitor.

a. A diode starts conduction only when its barrier potential is exceeded. For a Silicon diode, the barrier potential is 0.7 V. if the above diode is made with Silicon and the width of the depletion region is 3 μm, calculate the magnitude of the electric field intensity within the depletion region.

b. The dielectric constant of Silicon is 11.7. Given that, calculate the average electrostatic energy within the depletion region. The height of the depletion region is 7 μm and the depth of the diode is 10 μm.

c. Based on the answer you received in part a, for the electric field intensity and the permittivity of Silicon, calculate the magnitude of electric flux density.

d. The depletion region is a parallel plate capacitor. If one charge produces one flux line, based on the answer you received in part c and the surface area of the charge distribution; calculate the magnitude of total charge on one plate of the parallel plate capacitor.

e. Using your answer in part d, and the barrier voltage of 0.7 V, how much should be the capacitance within the depletion region?

9. Static charge eliminator for clean rooms.

Static charge buildup on surfaces can degrade the quality of textile, plastic, and printed products. This is also an issue in the medical industry since the glass apparatus used in the laboratories attracts dust due to the static charge buildup. To eliminate this issue, the industry uses static-charge eliminators. This question guides you along the process of static charge elimination.

a. Consider a slab of glass with dimensions 30 cm × 30 cm. If 10^{12} electrons are distributed on this glass slab, how much is the total charge in Coulombs?

b. If all the charges are uniformly distributed, calculate the surface charge density on the glass slab.

c. A static charge eliminator blows compressed air with a high concentration of protons or positive hydrogen ions. Protons have the same charge as an electron, but the sign is positive. The user can adjust the concentration and the blowing speed of compressed air. If the proton concentration of the static charge eliminator is 10000 per cubic centimeter. What is the total volume of compressed air needed to neutralize the static charge on the glass slab?

d. If the volume-exit velocity of the compressed air through the eliminator is ten cubic centimeters per second, how much of a current is produced at the mouth of the charge eliminator in Amperes?

10. Xerographic printing.
   a. Consider a printing drum with dimensions of a length of 30 cm, and a diameter of 3 cm.
   b. The diameter of a Selenium atom is 190 pm (pico-meters: 10^{-12}m). The Selenium atoms are distributed evenly only on the curved surface area of the drum. Approximately how many Selenium atoms can be distributed on the drum's curved surface area?
   c. If the thickness of the coating is 1000 Selenium atoms, how many Selenium atoms are used overall for the coating?
   d. The charge of Selenium is +2 (two protons). How much is the initial surface charge density on the drum in Coulombs?
   e. Assume the printer is printing the letter H with a total surface area of 2 cm². How much is the surface charge on the drum, on the area of the letter once it's imprinted on to the drum?
   f. The dry powder toner is made with negative resins with a charge of −1 (one electron) in the valance band. Approximately how many toner atoms will be attracted to the drum when printing the letter H?
   g. To print the toner on the paper, the paper should have a positive charge density twice the initial charge density of the drum. Determine the surface charge density that needs to be induced on a US letter-size paper (22 cm × 28 cm).

Magnetic Force, Energy, and Circuits

3

This chapter discusses the applications of magnetic forces, magnetic energy stored in components as well as magnetic circuits. The majority of the applications discussed in this section are based on Faraday's Law and the electromotive force. Faraday's Law is important in physics since it shows that a time-varying magnetic field can induce an electric field. In addition to Faraday's Law, another equation that is heavily used is Ampere's Law. This section discusses the displacement current term which was introduced by Maxwell to the Ampere's Law.

Application 15: Magnetic Shielding

Magnetic shielding is the process of warping or diverting the magnetic fields using high-permeability material. Magnetic permeability is the ability of a material to create an internal magnetic field when exposed to an external magnetic field. This makes the magnetic field pass through the highly permeable material—diverting from its original field.

Figure 3.1a shows the magnitude and direction of magnetic flux density on a copper plate from an electromagnet simulated on Ansys-Maxwell 3D. In this situation, the copper plate will be attracted to the electromagnet in the real world.

Figure 3.1b shows the magnetic field on a ferrite plate and the copper plate from the same electromagnet. In this case, the magnitude of the magnetic flux density on the copper plate is zero shown in dark blue. The reason is that the magnetic flux lines are compacted into ferrite hence no magnetic flux goes beyond ferrite.

A. Maxworth, *One Hundred Applications of Maxwell's Equations*, Synthesis Lectures on Electromagnetics, https://doi.org/10.1007/978-3-031-73784-8_3

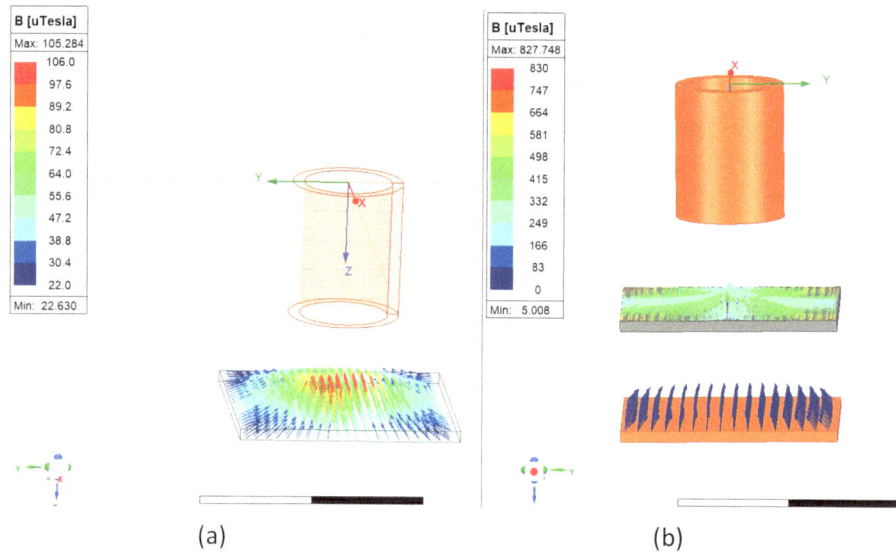

Fig. 3.1 a Magnetostatic simulation on Ansys Maxwell 3D of a copper plate underneath an electromagnet. **b** The copper plate and the electromagnet with a ferrite plate in between the two. Due to the ferrite plate the magnetic flux density on the copper plate is zero

Application 16: Calculating the Inductance of a Solenoid

The inductance of a material is its ability to generate an internal electromotive force or induce a voltage to oppose the direction of the supply voltage. Inductors are extensively used in electric circuits. The inductors are made from solenoids which are wounded wires. Hence, calculating the inductance of a solenoid is important in manufacturing inductors.

Imagine a scenario where a solenoid is connected to an alternating voltage. The current produced by this voltage source is also time-varying with a certain frequency. This time-varying current produces a time-varying magnetic field, and this time-varying magnetic field generates an electromotive force in the opposite direction of the supply voltage source in the circuit. The working principle is based on Faraday's law.

Faraday's law says: that when there is a time-varying magnetic field across a closed loop-it induces a voltage in the loop. This induced voltage is called the electromotive force since it can mobilize the electrons. The current generated due to this voltage or the EMF generates another magnetic field—opposite to the direction of the original magnetic field that induced the electromotive force. Hence the negative sign in Faraday's law. Faraday observed this dampening of the magnetic field through his experiments. But Lenz was the first person to introduce the negative sign in Faraday's Law. If the current generated due to EMF is in the same direction as the original magnetic field, the magnetic field keeps on increasing violating the Law of Energy Conservation.

Fig. 3.2 The cross section of the solenoid and the Amperian loop

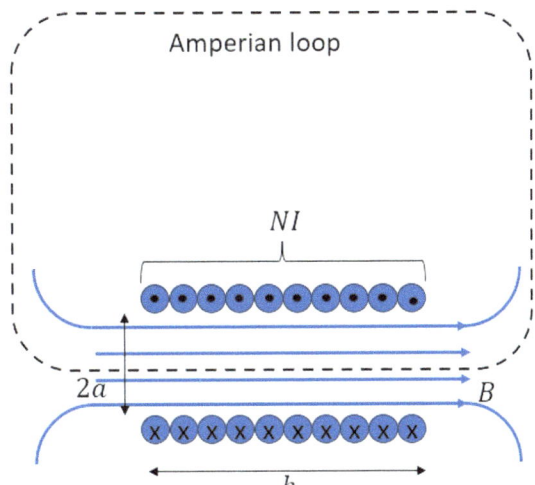

The inductance of one coil is called the self-inductance. When there are multiple coils present, we also must consider the mutual inductance between coils.

Let's consider the cross-section area of a coil, in which the current goes into the coil at the bottom of the windings and out of the coil at the top of the windings as shown in Fig. 3.2. Let's select a rectangular Amperian loop that goes through the axis of the coil. The edge of the rectangular path opposite to the edge going through the center of the coil is located further away from the loop such that there is no magnetic field along that edge. Let's assume that the axis of the coil is along the a_y.

The Ampere's law states that:

$$\oint_l \boldsymbol{H}.\boldsymbol{dl} = I_{enclosed} \tag{3.1}$$

Assuming that the coil is in the air:

$$\oint_l \boldsymbol{B}.\boldsymbol{dl} = \mu_o I_{enclosed} \tag{3.1a}$$

The only non-zero magnetic field exists inside the coil. The total current enclosed is *NI* Hence:

$$Bh = \mu_o NI \tag{3.1b}$$

$$\boldsymbol{B} = \frac{\mu_o NI}{h} \boldsymbol{a_y} \tag{3.1c}$$

The total magnetic flux inside the coil is:

$$\phi_m = \frac{\mu_o NI}{h}\left(\pi a^2\right) \tag{3.1d}$$

Inductance is the ratio between the magnetic flux and the current. Therefore, the inductance L:

$$L = \frac{\mu_o \pi a^2 N}{h} \tag{3.1e}$$

Application 17: Transformer EMF

The electromotive force has two forms: transformer EMF and motional EMF. Here, we talk about transformer EMF.

In transformer EMF, there is a stationary coil with an alternating current running through it. The alternating current generates a time-varying magnetic field.

Let's consider the transformer in the Fig. 3.3. The primary winding is connected to an alternating voltage source. This supply voltage source produces an alternating current that produces a time-varying magnetic field through the primary coil. This time-varying magnetic field increases in the direction of the current, hence the induced electromotive force or the transformer EMF opposes the direction of the current. If the magnetic flux goes through one loop of the primary coil is ϕ_m, the total EMF produced by the primary coil v_p is given by:

$$v_p = -N_p \frac{d\phi_m}{dt} \tag{3.2}$$

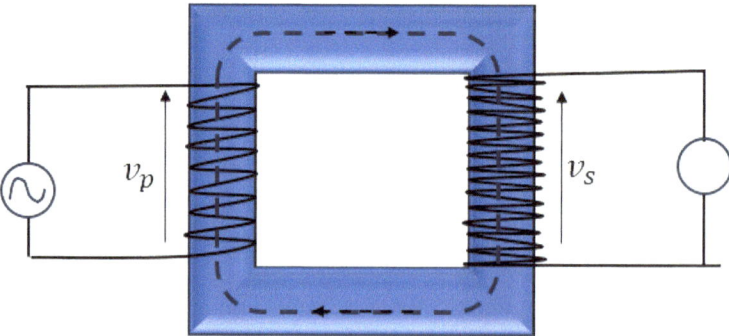

Fig. 3.3 Transformer EMF

In Eq. 3.2 N_p, is the number of windings in the primary coil. The primary and the secondary coils have no physical connection. But they are magnetically linked. Since the magnetic flux exists as loops, the magnetic flux runs through the core of the transformer. The core of a transformer is made with highly permeable materials such as ferrite since they can form internal magnetic flux lines.

If the secondary coil has N_s number of turns, the electromotive force induced at the secondary coil is:

$$v_s = -N_s \frac{d\phi_m}{dt} \tag{3.2a}$$

By dividing one equation from the other, we can get the transformer turn ratio:

$$\frac{v_p}{v_s} = \frac{N_p}{N_s} \tag{3.2b}$$

If $N_s > N_p$, it is called a step-up transformer, and if $N_s < N_p$, it's a step-down transformer.

Mutual Inductance

So far, we expressed the ability of a material to generate an electromotive force opposing the original increase in the magnetic field. This is called the self-inductance. When two inductors such as coils or solenoids are in the vicinity of each other in a magnetically coupled circuit, there can be a mutual inductance. Mutual inductance is the rate of change of magnetic flux in one solenoid, due to the current in the other. For example, the induced electromotive force on the secondary coil due to the current in the primary is:

$$\frac{d\phi_{sp}}{dt} = -M_{sp} \frac{di_p}{dt} \tag{3.2c}$$

Let's consider the circuit shown in Fig. 3.4. The inductance generated by the secondary coil from the current in the primary coil is defined as:

$$M_{sp} = \frac{N_s \phi_{sp}}{i_1} \tag{3.2d}$$

$$\phi_{sp} = BA = \frac{\mu N_p i_1 A}{l} \tag{3.2e}$$

In the above equation A is the cross-sectional area of the primary coil and l is the length of the coil and μ is the magnetic permeability of the core of the coil. Using the above two equations the inductance on the secondary coil due the current in the primary would be:

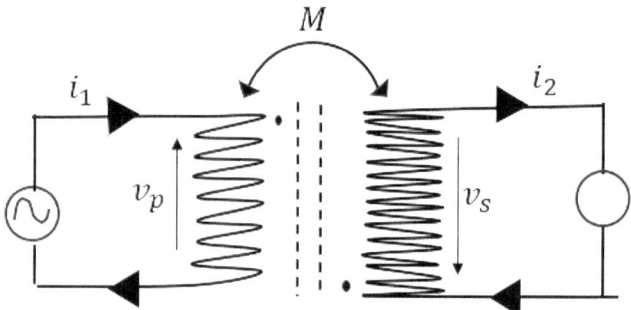

Fig. 3.4 The condition to add the mutual inductance to the voltage equations

$$M_{sp} = \frac{\mu N_p N_s A}{l} \tag{3.2f}$$

Similarly, the inductance on the primary coil due to the current in the secondary is:

$$M_{ps} = \frac{N_p \phi_{ps}}{i_2} \tag{3.2g}$$

Assuming both coils have the same cross-sectional area and the length, the magnetic flux becomes:

$$\phi_{ps} = \frac{\mu N_s i_2 A}{l} \tag{3.2h}$$

Therefore,

$$M_{ps} = \frac{\mu N_p N_s A}{l} \tag{3.2i}$$

Hence, $M_{sp} = M_{ps} = M$. Therefore, when analyzing transformer circuits, mutual inductance is indicated by M.

When analyzing the transformer as a circuit using mesh analysis, the impedance from the mutual inductance is either added or subtracted. The addition or subtraction depends on the windings of the transformer. In a transformer circuit symbol, the dot indicates the beginning of the windings. Figure 3.4 shows the condition where the mutual inductance should be added to the self-inductance based on the direction of the current and the position of the dot.

$$v_p = L_1 \frac{di_1}{dt} + M \frac{di_2}{dt} \tag{3.2j}$$

$$v_s = L_2 \frac{di_2}{dt} + M \frac{di_1}{dt} \tag{3.2k}$$

In the above equations L_1 and L_2 are the self-inductances of the primary and secondary windings.

Application 18: Magnetic Energy Stored in an Inductor

Analogous to electric energy, the magnetic energy can be defined as:

$$W_{magnetic} = \frac{1}{2} \int_v H.B dv = \frac{1}{2} \int_v \mu |H|^2 dv \qquad (3.3)$$

For a solenoid with N, the number of turns with a radius a and height h, and with a current i.

$$H = \frac{Ni}{h} a_z \qquad (3.3a)$$

$$W_{magnetic} = \frac{1}{2} \int_v \mu |H|^2 dv = \frac{1}{2} \mu \left(\frac{Ni}{h}\right)^2 \pi a^2 h \qquad (3.3b)$$

Let's rearrange the terms such that we can represent the magnetic energy in terms of the inductance and current.

$$W_{magnetic} = \frac{1}{2} \left(\frac{\mu N^2 \pi a^2}{h}\right) i^2 = \frac{1}{2} L i^2 \qquad (3.3c)$$

Application 19: The Electromagnet—Bell

An electric bell is one of the simplest applications of magnetic force and magnetic work. The Fig. 3.5 shows a simple electric bell used as a doorbell. The operation of the electric bell is as follows:

Step 1: when someone activates the switch, the electric circuit becomes complete and sends a current through the solenoid.

Step 2: the solenoid is wounded on a ferrous core and becomes an electromagnet.

Step 3: the electromagnet attracts the iron strip of the armature which is connected to a cast iron hammer (clapper).

Step 4: the lifting of the iron strip makes the hammer hit the cast iron gong (bell) making the sound.

Step 5: the lifting of the iron strip disconnects the circuit, stopping the power supply hence de-magnetizing the electromagnet—making the armature detach and come back to the initial position.

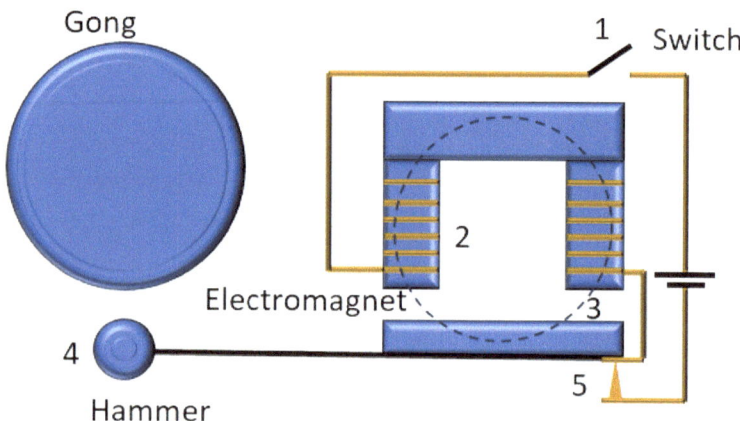

Fig. 3.5 The electric bell

Now, let's consider this circuit from the magnetic work perspective. Let's assume that the relative permeability of the ferrite core is μ_c, and the cross-sectional area of the electromagnet core is A. The lifting of the armature happens in the air. Let's assume that the lifting distance is Δl.

Ampere's law says that the total magnetic field around a closed path is:

$$\oint_l \mathbf{H}.\mathbf{dl} = I_{enclosed} \tag{3.4}$$

When we have N number of turns, the total current enclosed is:

$$\oint_l \mathbf{H}.\mathbf{dl} = NI = V_m \tag{3.4a}$$

The term V_m is known as the magneto-motive force.

The magnetomotive force is related to the magnetic flux ϕ_m, through the reluctance $\Re$ of the core.

The magnetic reluctance is analogous to electric resistance. Magnetic reluctance is defined as the ratio between the magneto-motive force (MMF) to magnetic flux. Like electric resistance, magnetic reluctance is proportional to the mean path length of the magnetic material and inversely proportional to its cross-sectional area. The proportionality constant is the reciprocal of magnetic permeability—the higher the magnetic permeability, the lower the magnetic reluctance. The equation below shows the magnetic reluctance of the ferrite core. l is the mean path length of the core, and A is the cross-sectional area.

$$\mathfrak{R}_{core} = \frac{l_{core}}{\mu_c \mu_o A} \tag{3.4b}$$

Since the above, circuit has an airgap, the reluctance of air is:

$$\mathfrak{R}_{air} = \frac{l_{air}}{\mu_o A} \tag{3.4c}$$

The magnetic flux

$$\phi_m = \frac{V_m}{\mathfrak{R}_{core} + \mathfrak{R}_{air}} \tag{3.4d}$$

Remember that the magnetic flux is continuous throughout the circuit. Let's define the parameters, B_{core} the magnitude of the magnetic flux density inside the core, B_{air}, the magnitude of the magnetic flux density in air, H_{core}, the magnitude of magnetic field intensity in the core, and H_{air}, the magnitude of magnetic field intensity in air.

$$\phi_m = B_{core} A \tag{3.4e}$$

Since the magnetic flux is continuous, and the cross-section of the core where the flux leaks into air is A:

$$\phi_m = B_{air} A \tag{3.4f}$$

This leads, to the relationship:

$$B_{core} = B_{air} \tag{3.4g}$$

$$\mu_c \mu_o H_{core} = \mu_o H_{air} \tag{3.4h}$$

$$H_{core} = \frac{H_{air}}{\mu_c} \tag{3.4i}$$

The above relationships show that although the magnetic flux density of the core and air are the same, the magnetic field intensity of air is μ_c, times higher than the magnetic field intensity of the core. This high-intensity magnetic field is important in all magnetic circuit applications where electromagnets are used. Now, let's find out the magnetic force and magnetic work done in a bell.

As we saw in the previous example, the magnetic work is:

$$W_{magnetic} = \frac{1}{2} \int_v \boldsymbol{H}.\boldsymbol{B} dv = \frac{1}{2} \int_v \mu |\boldsymbol{H}|^2 dv \tag{3.4j}$$

In the case of the doorbell, the armature lifting (work) happens in the air. Therefore:

$$W_{magnetic} = \frac{1}{2} \int_v \mu_o H_{air}^2 dv \tag{3.4k}$$

The total volume where the work is being done is: $2A\Delta l$

Therefore, the total average magnetic work done in a bell with an electromagnet is:

$$W_{magnetic} = \frac{1}{2} \mu_o H_{air}^2 2A\Delta l = \mu_o H_{air}^2 A\Delta l \tag{3.4l}$$

The magnetic force is related to magnetic work as:

$$W_{magnetic} = \boldsymbol{F_m}.\Delta \boldsymbol{l} \tag{3.4m}$$

Hence the magnitude of the magnetic force $\boldsymbol{F_m} = \mu_o \frac{H_{air}^2 A\Delta l}{\Delta l}$

$$|\boldsymbol{F_m}| = \mu_o H_{air}^2 A \tag{3.4n}$$

Application 20: Magnetic Valves

Magnetic valves are another application of magnetic circuits. Magnetic valves are widely used in fluid control applications. The size of the magnetic valves varies with the application. The magnetic valves can be categorized into two: normally closed, and normally open. In both cases, the valve system consists of an electromagnet, a plunger, and a spring.

Let's consider the normally closed magnetic valve first. The Fig. 3.6a shows the cross-section of the coil encircling the plunger and the spring. When the circuit is activated, the current running in the coil generates a magnetic field. This magnetic field is the strongest at the center of the coil. The direction of the magnetic field is given by the right-hand rule—*when we wrap the fingers of the right hand around the coil (solenoid) such that the thumb is pointed in the direction of the conventional current, the thumb points to the north pole and the other fingers point to the direction of the current through the solenoid.* In the normally closed magnetic valve, the goal is to lift the plunger when the coil is energized, thence the magnetic field inside the coil should be upwards, and the arrow shows the direction of the conventional current. Once the current to the coil is turned off, the plunger is brought to its original position by the spring.

In the case of the normally open magnetic valve shown in Fig. 3.6b, the only thing that changes is the direction of the conventional current through the coil. In this case, the plunger should be pushed downwards by the magnetic field when the circuit is turned on. Hence the magnetic field inside the coil should be downwards. Once the current to the coil is turned off, the spring brings the plunger back to its original open position.

Let's calculate the magnetic force applied to the plunger in this case. Remember, if the dimensions and the material stay the same, the only thing changes between the normally

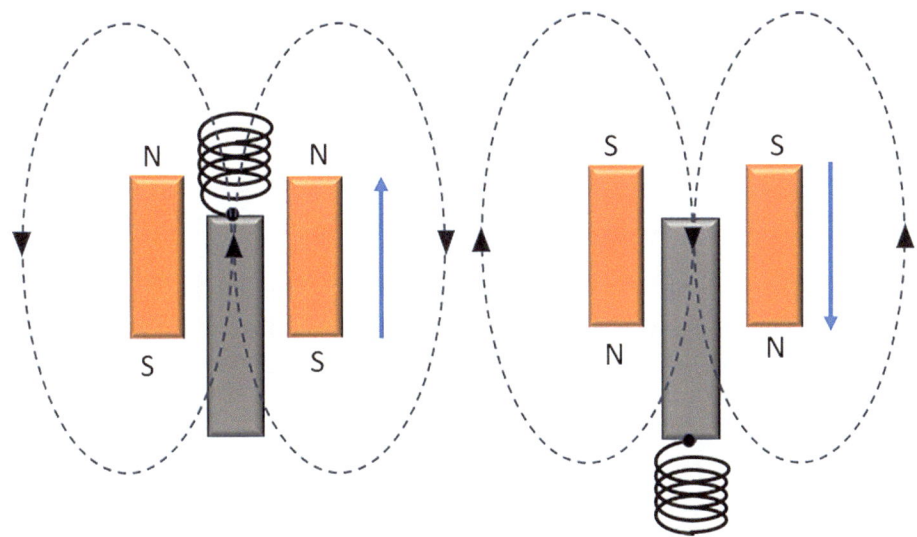

Fig. 3.6 a The normally closed magnetic valve. **b** The normally open magnetic valve

closed, and normally open valves is the direction of the force. Assuming that the entire system is in the air, according the Ampere's law:

$$\oint_l \boldsymbol{B}.\boldsymbol{dl} = \mu_o I_{enclosed} \tag{3.5}$$

As we found out in the inductor example, the magnitude of the magnetic flux density along the center axis of the coil with N turns, and length h is:

$$B = \frac{\mu_o NI}{h} \tag{3.5a}$$

If the coil radius is a, the total magnetic flux passing through the coil is:

$$\phi_m = \frac{\mu_o NI}{h}\left(\pi a^2\right) \tag{3.5b}$$

Now, let's assume that the cross-section of the plunger is A, and the plunger is either lifted or pushed down by a distance of Δl.

The average magnetic work:

$$W_{magnetic} = \frac{1}{2}\mu_o H_{air}{}^2 A\Delta l = \frac{1}{2}\mu_o \left(\frac{NI}{h}\right)^2 A\Delta l \tag{3.5c}$$

Note that $H_{air} = \frac{B}{\mu_o} = \frac{NI}{h}$.

The magnitude of the magnetic force applied on the plunger is:

$$|\boldsymbol{F_m}| = \frac{1}{2}\mu_o\left(\frac{NI}{h}\right)^2 A \tag{3.5d}$$

Application 21: Electromagnetic Levitation—Levitating Globe

The levitating globe is the simplest example of magnetic levitation. Magnetic levitation is based on the preliminary principles of magnetism: opposite magnetic poles attract and the like poles repel. Figure 3.7 shows the electromagnetic circuitry inside a levitation globe and a base. The arrangement and the working principles of the levitating globe can be broken down into the following steps:

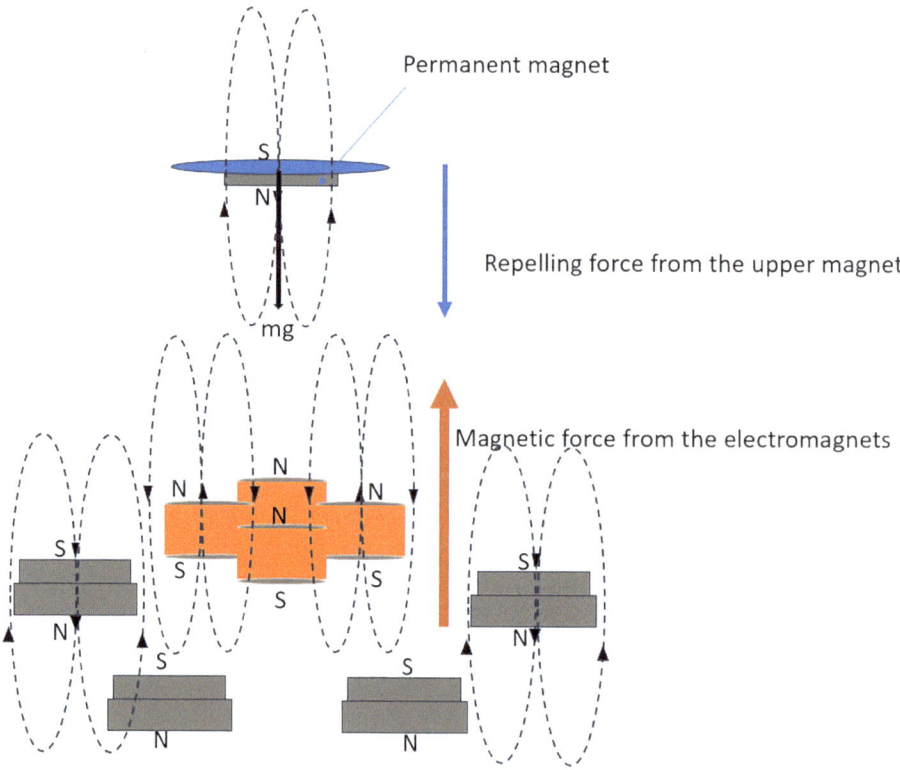

Fig. 3.7 Direction of magnetic fields produced by the permanent and electromagnets inside a levitating globe

- The base is comprised of four electromagnets and seven permanent magnets.
- The current through the electromagnets is such that the north pole faces upwards. The permanent magnets are arranged such that their south pole faces upwards. (Remember the right-hand thumb rule: when you wrap your right hand around a current-carrying solenoid with the thumb pointed in the direction of conventional current, the fingers curl in the direction of the current through the solenoid, and the thumb points to the north pole.)
- Inside the levitating globe, there is another permanent magnet with its north pole facing down.
- Given that like poles repel, the north pole from the magnet inside the levitating globe is repelled by the magnetic force from the electromagnets. In other words, the electromagnets are the ones keeping the globe levitating. The electromagnets generate enough magnetic force to repel the magnetic force from the permanent magnet inside the globe and to oppose the weight of the globe.
- Each electromagnet is connected to a Hall sensor. Hall sensors automatically switch on when they detect a magnetic field above a threshold. When the globe is not centered the magnetic force detected by the four Hall sensors will not be equal and some sensors will switch on. Hence it will command the control system to increase the current through the electromagnets handling more weight and more repelling force. This stronger force from the electromagnets will force the globe back to the center.
- The surrounding south-facing permanent magnets in the base act as a protective fence. If the globe falls off due to a malfunctioning of the electromagnets, it will quickly get attracted to the south-facing permanent magnets in the base.

References

- Website: https://www.kjmagnetics.com/blog.asp?p=electromagnetic-levitation accessed June 25, 2024.
- Website: https://www.youtube.com/watch?v=eOT_G-1ogn4 accessed June 25, 2024.

Application 22: Electromagnetic Crane

Electromagnetic cranes are used in heavy cargo lifting at docks as well as in junk yards. These heavy-duty cranes can lift metallic waste from dump sites to lifting container cargo at the shipyards. The operational principle of an electromagnetic crane is the same as the magnetic circuits that we discussed above—the only difference being the strength of the magnetic force and the magnetic work done. These magnets have a large number of turns and a high current running through them to produce the lifting force. The Fig. 3.8 shows a cross-section of the electromagnet in a crane. The iron core of the electromagnet has a shape of an inverted W shape to increase the cross-sectional area of the core and hence

Fig. 3.8 Cross sectional area of the electromagnet inside a crane

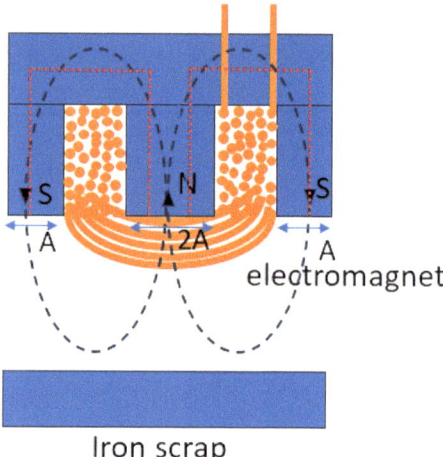

the magnetic field strength. The current through the coil is such that the center of the iron core acts as the north pole of the electromagnet. The iron core of the north pole is double the size of the two south poles. The magnet is designed such that the magnetic flux at the center of the magnet is upwards—producing an upward force.

Let's follow a similar process as we did for the electromagnetic bell and calculate the work done when the electromagnetic crane is lifting a piece of heavy metal. Starting from Ampere's law which says, the total electromagnetic field intensity along a closed path equals the current enclosed.

$$\oint_l \boldsymbol{H}.d\boldsymbol{l} = I_{enclosed} \tag{3.6}$$

From that, we get the magnetomotive force, which is the number of turns in a coil, times the current.

$$\oint_l \boldsymbol{H}.d\boldsymbol{l} = NI = V_m \tag{3.6a}$$

The reluctance of the core and the air are given as follows. In the case of the electromagnetic crane, the magnetic flux emitted by the north pole is split into two equal portions. Therefore, it will be easier to analyze this system as two magnetic circuits are kept back-to-back.

$$\mathfrak{R}_{core} = \frac{l_{core}}{\mu_c \mu_o A} \tag{3.6b}$$

$$\mathfrak{R}_{air} = \frac{l_{air}}{\mu_o A} \tag{3.6c}$$

$$\phi_m = \frac{V_m}{\mathfrak{R}_{core} + \mathfrak{R}_{air}} \tag{3.6d}$$

The total magnetic flux is twice the magnetic flux produced by one circuit. Hence the total magnetic flux is:

$$\phi_{total} = \frac{2V_m}{\mathfrak{R}_{core} + \mathfrak{R}_{air}} \tag{3.6e}$$

The magnetic flux is continuous; hence the magnetic flux densities are the same for both air and the core. $B_{core} = B_{air}$.

$$H_{core} = \frac{H_{air}}{\mu_c} \tag{3.6f}$$

$$W_{magnetic} = \frac{1}{2} \int_v \mu_o H_{air}^2 dv \tag{3.6g}$$

$$W_{magnetic} = \frac{1}{2} \mu_o H_{air}^2 4A\Delta l = 2\mu_o H_{air}^2 A\Delta l \tag{3.6h}$$

In Eq. 3.6h, A is the cross-section area of the iron core south pole. The north pole cross-section is twice that.

Using the relationship between the magnetic work and the force, the magnetic pull force can be calculated as:

$$|\boldsymbol{F_m}| = 2\mu_o H_{air}^2 A \tag{3.6i}$$

Reference

- Website: https://www.rfcafe.com/references/Electricity-Basic-Navy-Training-Courses/electricity-basic-navy-training-courses-chapter-12.htm accessed June 25, 2024.

Application 23: Moving Coil Speaker

The purpose of a speaker is to convert electric energy to acoustic energy. A moving coil speaker shown in Fig. 3.9 consists of a thin-wire copper coil that acts as an electromagnet wound on a heat-resistant cylinder called a former, a permanent magnet with two steel plates at the top and the bottom. This coil and the magnet system are called the motor of the speaker. In addition to these parts, there is a cone connected to a corrugated spider.

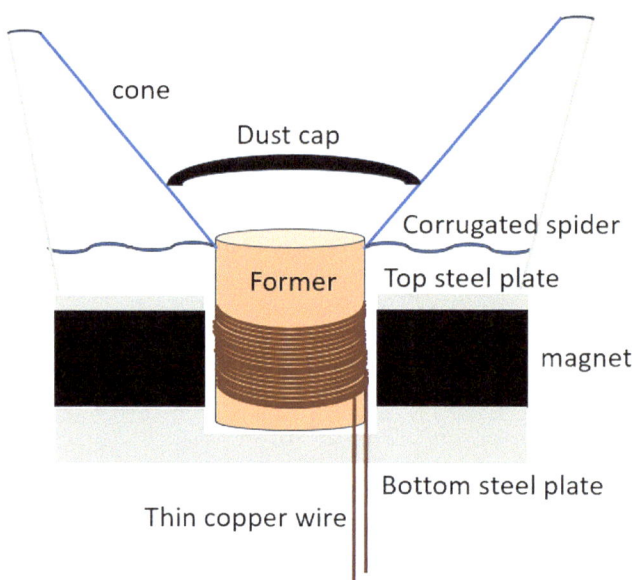

Fig. 3.9 A moving coil speaker

The cone and the corrugated system are connected to the coil which works as an electromagnet. The cone is suspended at the edge so it can move back and forth laterally. A typical speaker also has a dust cap to protect the interior from dust.

The coil receives current from the audio amplifier. When there is a current through the coil, it produces a magnetic field. The strength and the direction of this magnetic field is proportional to the amplitude of the current. Based on the strength and the magnetic field of the electromagnet, the coil gets either attracted or repelled by the permanent magnet. As the coil moves, the suspended cone is also pulled or pushed with it. This movement of the cone pulls or pushes air molecules creating pressure variations that our ear perceives as sound waves.

As the amplitude of the sound signal varies with the amplitude of the current through the electromagnet, the frequency of the signal varies with the frequency of the current. As the frequency of the current changes, the frequency of the magnetic field through the electromagnet changes, changing the pulling and pushing frequency of the cone. Which in turn, generates pressure waves in that frequency that we hear as the frequency or the pitch of the sound signal.

Reference

- Website: https://www.youtube.com/watch?v=RxdFP31QYAg accessed June 25, 2024.

Application 24: Electric Relay—Interrupter—Miniature Circuit Breaker

The miniature circuit breaker shown in Fig. 3.10a is a mechanical relay circuit, where the electromagnet is energized when there is an overload or a short circuit, pulling the plunger of the trip bar toward it (3.10b) Under normal current conditions shown in Fig. 3.10a, the force produced by the electromagnet is not strong enough to pull the plunger. When there is a short circuit within the system, there is a heavy current going through the circuitry, and the electromagnet produces a larger force than usual—which is strong enough to pull the plunger. Once the plunger is pulled towards the electromagnet, the latch system opens the switch disconnecting the current. A standard miniature circuit breaker will turn off the current through the system within 2.5 ms of encountering an overload.

The purpose of the bimetallic strip is to detect overheating conditions. The bimetallic strip is made of two metals with different thermal properties. When an overheating condition is encountered the two metals expand unevenly—causing the bimetallic strip to bend. Once the bent bimetallic strip touches the trip bar, this pushes the plunger towards the electromagnet—making the latch system open the switch. When there is a short circuit scenario, there will be a current overload, and this overload can generate extra heat within the system. Therefore, in a miniature circuit breaker, the pulling of the plunger from the electromagnet and the thermal expansion of the bimetallic strip happen simultaneously.

References

- Website: https://electrical-engineering-portal.com/miniature-circuit-breakers-mcbs-for-beginners accessed June 25, 2024.
- Website: https://www.engineersgarage.com/insight-how-mcb-works/ accessed June 25, 2024.

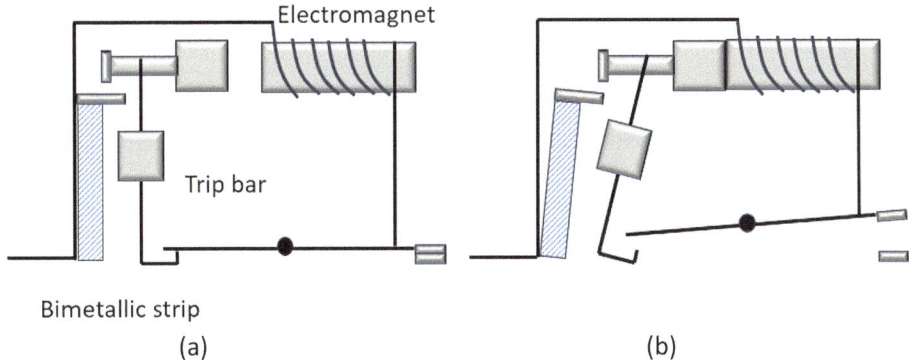

Fig. 3.10 **a** The electric relay under normal current condition. **b** The electric relay under overload conditions

Application 25: Magnetic Measurement of Coating Thickness

The magnetic measurement of coating thickness works based on the force required to pull a magnet out of a ferromagnetic surface. This method can be used particularly in measuring the coating thickness of paint on highly permeable material. The derivation here is based on the fact that the force required to pull the magnet off the coated surface is the same as the force required to pull the coated surface toward the magnet.

Let's consider an electromagnet with N turns with a current of I. The length of the solenoid is L. From Ampere's law, we can get that the average magnetic field intensity along the axis of the solenoid is:

$$H = \frac{NI}{L} \tag{3.7}$$

When the solenoid is wound around a ferromagnetic material with relative permeability μ_c, the magnitude of the magnetic flux density inside the ferromagnetic core is:

$$B_c = \mu_c \mu_o \frac{NI}{L} \tag{3.7a}$$

Given that the magnetic flux lines are continuous, the magnetic flux density inside the ferromagnetic core of the electromagnet and air are the same. If this magnet is used to lift a piece of iron in the air is:

$$W_{magnetic} = \frac{1}{2} \int_v \boldsymbol{H_{air}} . \boldsymbol{B_{air}} dv \tag{3.7b}$$

$$W_{magnetic} = \frac{1}{2} \frac{B_c^2}{\mu_o} A \Delta l \tag{3.7c}$$

In the above equation, A is the area of the electromagnet, and Δl is the lifting length. Which brings us to the force equation,

$$F = \frac{1}{2} \frac{B_c^2}{\mu_o} A \tag{3.7d}$$

When the material is coated with coating thickness t, the work required to lift the material reduces to:

$$W'_{magnetic} = \frac{1}{2} \frac{B_c^2}{\mu_o} A(\Delta l - t) \tag{3.7e}$$

$$W'_{magnetic} = \frac{1}{2} \frac{B_c^2}{\mu_o} A \Delta l \left(1 - \frac{t}{\Delta l} \right) \tag{3.7f}$$

The difference in magnetic work when lifting a coated vs un-coated permeable material is:

$$\frac{W'_{magnetic}}{W_{magnetic}} = \frac{\Delta l - t}{\Delta l} \tag{3.7g}$$

The new force required to lift a coated surface is:

$$F_{coated} = F\left(1 - \frac{t}{\Delta l}\right) \tag{3.7h}$$

In Eq. 3.7h F is the force required to lift the highly permeable material a distance of Δl.

Application 26: Inductance of a Coaxial Cable

The inductance of a coaxial cable can be calculated as follows. Let's consider the radius of the conductor as a, and the inner radius of the outer copper mesh as b. And let's consider the current through the inner conductor is I. The axis of the coaxial cable is along the Z axis. $\mu(= \mu_r\mu_o)$ is the permeability of the low-loss dielectric between the two conductors.

From the Ampere's Law:

$$\oint_l \mathbf{H}.d\mathbf{l} = I_{enclosed} \tag{3.8}$$

$$\oint_l \mathbf{B}.d\mathbf{l} = \mu I_{enclosed} \tag{3.8a}$$

From the right-hand rule, for a current in the $\mathbf{a}_z$ direction, the magnetic field will be in the positive $\mathbf{a}_\phi$ direction. And the infinitesimal path length would be $\rho d\phi$ in the $\mathbf{a}_\phi$ direction.

Hence the integration along the path will yield:

$$\int_0^{2\pi} B_\phi \mathbf{a}_\phi.\rho d\phi \mathbf{a}_\phi = \mu I \tag{3.8b}$$

$$B_\phi = \frac{\mu I}{2\pi\rho} \tag{3.8c}$$

To find the inductance we need the total magnetic flux stored within the region between the two conductors. For that, we need to integrate the magnetic flux density over an area, such that the area of integration cuts the magnetic flux perpendicularly. The figure below shows the coaxial cable arrangement together with the area of choice that is used for the integration. Another way to approach this problem is that the magnetic flux density in this case is in the a_ϕ, direction. Hence, the area should also be in the same direction (remember when it comes to infinitesimal area elements, the direction of the area is perpendicular to the physical surface area).

The a_ϕ directed infinitesimal area is $d\rho dz a_\phi$. The integration becomes:

$$\int_{\rho=a}^{b} \int_{z=0}^{h} \frac{\mu I}{2\pi\rho} a_\phi . d\rho dz a_\phi = \phi_m \tag{3.8d}$$

$$\Phi_m = \frac{\mu I h}{2\pi} \{\ln(b) - \ln(a)\} \tag{3.8e}$$

$$\Phi_m = \frac{\mu I h}{2\pi} \ln\left(\frac{b}{a}\right) \tag{3.8f}$$

Inductance is the magnetic flux per unit current. Therefore,

$$L = \frac{\Phi_m}{I} = \frac{\mu h}{2\pi} \ln\left(\frac{b}{a}\right) \tag{3.8g}$$

Also, let's define the inductance per unit length of a coaxial cable:

$$L' = \frac{\mu}{2\pi} \ln\left(\frac{b}{a}\right) \tag{3.8h}$$

Application 27: Displacement Current

The invention of the displacement current is the main reason why all these four laws Gauss's, Faraday's, and Ampere's laws are called Maxwell's equations. The introduction of the displacement current to Ampere's law showed the missing link between the electromagnetic fields. Faraday's law shows that a time-varying magnetic field can generate a time-varying electric field. The displacement current term introduced to Ampere's law by Maxwell showed that a time-varying electric field can in-tern generate a time-varying magnetic field, hence the birth of electromagnetic fields.

To demonstrate the importance of the displacement currents, let's consider the simple R–C circuit connected to a current source. The circuit current is $I_o e^{i\omega t}$. The basic Ampere's law says the magnetic field along is closed loop is equal to the current enclosed. In this circuit, the current through the conductive paths is equal to $I_o e^{i\omega t}$. But how about inside the capacitor? According to the basic Ampere's law, the current through the capacitor is zero since there are no conductors inside. But the current cannot disappear suddenly and reappear—it must be continuous throughout the circuit. Hence the importance of the displacement current. The displacement current I_D is defined as the rate of change of electric flux.

Now let's show that the current is continuous and the displacement current through the capacitor is equal to the circuit current. Let's define the parameter J_d, which is the displacement current density. The circuit current accumulates charges on the capacitor plates initiating the displacement currents. Although a vector, here let's consider only the magnitude of the displacement current density. The direction of the displacement current density is the same as the direction of the circuit current.

$$J_d(t) = \frac{I_o e^{i\omega t}}{S} \tag{3.9}$$

According to Gauss's law, electric flux is proportional to the charge enclosed.

$$\phi_e = Q(t) \tag{3.9a}$$

$$\frac{d\phi_e}{dt} = \frac{dQ(t)}{dt} \tag{3.9b}$$

$$\frac{d\phi_e}{dt} = \frac{d}{dt}\left(S\int_t J_d(t).dt\right) \tag{3.9c}$$

The above operation gives:

$$\frac{d\phi_e}{dt} = I_o e^{i\omega t} \tag{3.9d}$$

Which is the same as the circuital current. This shows that the current is continuous throughout the circuit.

Application 28: Kirchoff's Voltage Law

Kirchoff's voltage law states that the sum of voltages across a closed loop is zero and it's being widely used in electrical circuit analysis. The Kirchhoff's voltage law can be derived using the Faraday's Law. On the right-hand side of Faraday's law, if the rate of change of magnetic flux through the loop is zero, the integration of electric field intensity over the path (which is electric potential) will be zero. This is true for direct current circuits. Since direct currents generate a steady state magnetic field around the conductors, as long as the loop stays steady, the rate of change of magnetic flux through the loop will be zero. Hence Kirchoff's voltage law is valid.

$$\oint_l \mathbf{E}.d\mathbf{l} = -\frac{\partial \phi_m}{\partial t} \tag{3.10}$$

For steady-state circuits $\frac{\partial \phi_m}{\partial t} = 0$. When the circuit has discrete elements, the voltage across the entire loop can be calculated using the summation:

$$\sum_{i=1}^{n} V_i = 0 \tag{3.10a}$$

Figure 3.11 shows a circuit consists of a DC voltage source and two resistors created using National Instrument Multisim software. The arrows point to the high potential. As shown in the figure, the source voltage is equal to the sum of voltages across the two resistors. The net voltage across the loop is zero.

But how about with alternating current courses? Alternating currents are time-varying hence, they generate time-varying magnetic fields. Hence according to Faraday's law, there is a non-zero electromotive force. Is the Kirchoff's law still valid? The short answer is no—it's not. When dealing with alternating currents, there is an electromotive voltage generated due to the time-varying magnetic field through the entire circuit loop (i.e. a closed circuit is a loop). This electromotive force through the entire circuit is often

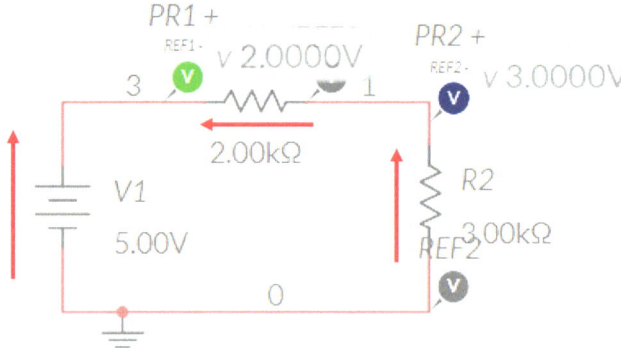

Fig. 3.11 An electrical circuit composed of a DC voltage source, two resistors connected in series and two differential (difference) probes

neglected in circuit theory. In many textbook explanations, it is given that Kirchoff's voltage law is applicable only instantaneously—which means at any given time point Kirchoff's voltage law holds.

When there is an inductor is present as a circuit element, the Faraday's law equation for the circuit is:

$$v_s(t) - v_R(t) - v_c(t) = -\frac{d\phi_m}{dt} \tag{3.10b}$$

Now, let's write the magnetic flux in terms of inductance and current. Remember, the inductance is the magnetic flux created by a unit current, in the opposite direction of the original magnetic flux increase from the source. Hence:

$$v_s(t) - v_R(t) - v_c(t) = -\frac{d(-Li(t))}{dt} \tag{3.10c}$$

To get the differential equation for the circuit, let's use the other relationships such as $v_c(t) = \frac{Q(t)}{C}$ and $i(t) = \frac{dQ(t)}{dt}$, hence $q(t) = \int_t i(t).dt$

$$v_s(t) - Ri(t) - \frac{1}{C}\int_t i(t).dt = L\frac{d(i(t))}{dt} \tag{3.10d}$$

Application 29: Kirchoff's Current Law

Kirchhoff's current law says the net current at a node of an electric circuit is equal to zero. In other words, the current coming into the node equals the current going out of the node. Like Kirchhoff's voltage law, Kirchhoff's current law is also derived from Maxwell's equations. In this case, let us start with Ampere's Law with Maxwell's contribution of displacement current.

$$\nabla \times \boldsymbol{H} = \sigma \boldsymbol{E} + \frac{\partial \boldsymbol{D}}{\partial t} \tag{3.11a}$$

$$\boldsymbol{J} = \sigma \boldsymbol{E} \tag{3.11b}$$

Let us consider the vector identity that the divergence of a curl of a vector field equals zero.

$$\nabla.(\nabla \times \boldsymbol{H}) = \nabla.\left(\boldsymbol{J} + \frac{\partial \boldsymbol{D}}{\partial t}\right) = 0 \tag{3.11c}$$

$$\nabla.\boldsymbol{J} + \nabla.\frac{\partial \boldsymbol{D}}{\partial t} = 0 \tag{3.11d}$$

Now, let us consider the node as a conductive volume. Within a conductor, the variation of electric flux is zero. Hence:

$$\oint_v \nabla.\boldsymbol{J}\,dv = 0 \tag{3.11e}$$

Using the divergence theorem, we can convert the above volume integration to a surface integration.

$$\oint_v \nabla.\boldsymbol{J}\,dv = \oint_s \boldsymbol{J}.d\boldsymbol{s} = 0 \tag{3.11f}$$

The above equation means the total current at the surface of the node is zero. Hence:

$$\sum_{i=1}^{n} I_i = 0 \tag{3.11g}$$

Figure 3.12 shows a circuit consists of a current source and three resistors simulated using National Instrument Multisim software. According to the simulation the 3A current coming into the node is split into a 2A and 1A currents. The algebraic sum of currents at the node is zero agreeing with the Kirchoff's current law.

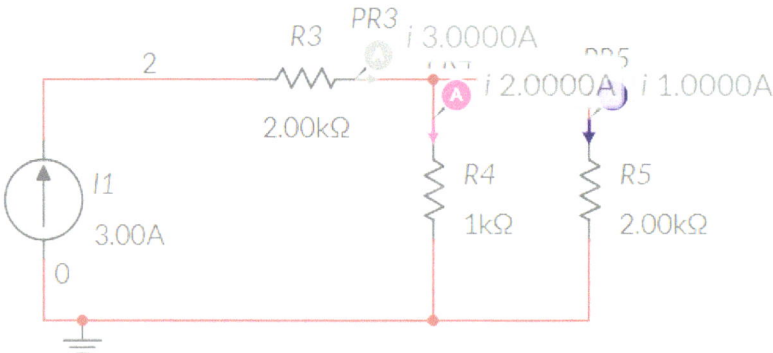

Fig. 3.12 An electric circuit with a direct current source, three resistors and three current probes simulated using NI Multisim software

Application 30: Derivation of the Biot-Savart Law

Biot-Savart Law named after French physicists Jean Baptiste Biot and Felix Savart is a fundamental law in magnetostatics. Biot-Savart law is extensively used in determining the magnitude and direction of the magnetic fields with loop currents. This law has its applications in analyzing loop antennas and ring currents. The Bio-Savart Law is derived from the Ampere's law and the Gauss's law for magnetic fields. Hence the Biot-Savart Law can be applied to scenarios where Ampere's law is not applicable.

Here is the derivation of the Biot-Savart law:

$$\nabla.\boldsymbol{B} = 0 \tag{3.12a}$$

$$\nabla \times \boldsymbol{B} = \mu \boldsymbol{J} \tag{3.12b}$$

Using the vector identity a non-diverging field is rotational. We can express the magnetic flux density using the magnetic vector potential $\boldsymbol{A}$.

$$\boldsymbol{B} = \nabla \times \boldsymbol{A} \tag{3.12c}$$

$$\nabla \times (\nabla \times \boldsymbol{A}) = \nabla(\nabla.\boldsymbol{A}) - \nabla^2\boldsymbol{A} \tag{3.12d}$$

$$\nabla \times \boldsymbol{B} = -\nabla^2\boldsymbol{A} \tag{3.12e}$$

$\nabla.\boldsymbol{A} = 0$ is known as the Coulomb's gauge condition.

$$\nabla^2\boldsymbol{A} = -\mu \boldsymbol{J} \tag{3.12f}$$

Given that both $\boldsymbol{A}$ and $\boldsymbol{J}$ are vectors:

$$\nabla^2\boldsymbol{A}_x = -\mu \boldsymbol{J}_x \tag{3.12g}$$

$$\nabla^2\boldsymbol{A}_y = -\mu \boldsymbol{J}_y \tag{3.12h}$$

$$\nabla^2\boldsymbol{A}_z = -\mu \boldsymbol{J}_z \tag{3.12i}$$

At this point, let's use the Green's theorem. Green's theorem says that Green's function is the response when a linear operator is applied to a Dirac-delta function. In the above equations, the Laplacian operator is the linear operator, the vector magnetic potential is the response, and the current densities are the sources. Equations 3.12j and 3.12k represent the position vectors of the observer and the source.

$$r = xa_x + ya_y + za_z \tag{3.12j}$$

$$r' = x'a_x + y'^{a_y} + z'a_z \tag{3.12k}$$

Let

$$R = r - r' = (x - x')a_x + (y - y')a_y + (z - z')a_x \tag{3.12l}$$

$$R = |r - r\prime| = \sqrt{(x - x')^2 + (y - y')^2 + (z - z')^2} \tag{3.12m}$$

The Green's function for the Laplacian operator is the well-known:

$$G(r, r') = -\frac{1}{4\pi |r - r'|} \tag{3.12n}$$

In the above Green's function, r is the observer location, and r' is the source location. The convolution between the Green's function and the source functions should provide the responses required. Hence:

$$A_x = \frac{\mu}{4\pi} \int_v \frac{J_x(r')}{|r - r'|} dv \tag{3.12o}$$

$$A_y = \frac{\mu}{4\pi} \int_v \frac{J_y(r\prime)}{|r - r\prime|} dv \tag{3.12p}$$

$$A_z = \frac{\mu}{4\pi} \int_v \frac{J_z(r\prime)}{|r - r'|} dv \tag{3.12q}$$

and

$$B = \nabla \times A \tag{3.12r}$$

$$B = \frac{\mu}{4\pi} \int_v \nabla \times \frac{J(r')}{|r - r'|} dv \tag{3.12s}$$

Let's derive one component of the magnetic field:

$$B_x = \frac{\partial}{\partial y} A_z - \frac{\partial}{\partial z} A_y \tag{3.12t}$$

$$B_y = \frac{\partial}{\partial z} A_x - \frac{\partial}{\partial x} A_z \tag{3.12u}$$

$$B_z = \frac{\partial}{\partial x}A_y - \frac{\partial}{\partial y}A_x \tag{3.12v}$$

$$B_x = \frac{\mu}{4\pi}\int_v J_z(r')\frac{\partial}{\partial y}\left(\frac{1}{|r-r'|}\right) - J_y(r')\frac{\partial}{\partial z}\left(\frac{1}{|r-r'|}\right)dv \tag{3.12w}$$

$$B_y = \frac{\mu}{4\pi}\int_v J_x(r')\frac{\partial}{\partial z}\left(\frac{1}{|r-r'|}\right) - J_z(r')\frac{\partial}{\partial x}\left(\frac{1}{|r-r'|}\right)dv \tag{3.12x}$$

$$B_z = \frac{\mu}{4\pi}\int_v J_y(r')\frac{\partial}{\partial x}\left(\frac{1}{|r-r'|}\right) - J_x(r')\frac{\partial}{\partial y}\left(\frac{1}{|r-r'|}\right)dv \tag{3.12y}$$

$$\frac{\partial}{\partial y}\left(\frac{1}{|r-r'|}\right) = -\frac{1}{2}\frac{2(y-y')}{\left(\sqrt{(x-x')^2+(y-y')^2+(z-z')^2}\right)^3} \tag{3.12z}$$

$$\frac{\partial}{\partial x}\left(\frac{1}{|r-r'|}\right) = -\frac{(x-x')}{|r-r'|^3} \tag{3.12aa}$$

$$\frac{\partial}{\partial y}\left(\frac{1}{|r-r'|}\right) = -\frac{(y-y')}{|r-r'|^3} \tag{3.12ab}$$

$$\frac{\partial}{\partial z}\left(\frac{1}{|r-r'|}\right) = -\frac{(z-z')}{|r-r'|^3} \tag{3.12ac}$$

$$B_x = \frac{\mu}{4\pi}\int_v J_y(r')\frac{R_z}{R^3} - J_z(r')\frac{R_y}{R^3}dv \tag{3.12ad}$$

$$B_y = \frac{\mu}{4\pi}\int_v J_z(r')\frac{R_x}{R^3} - J_x(r')\frac{R_z}{R^3}dv \tag{3.12ae}$$

$$B_z = \frac{\mu}{4\pi}\int_v J_x(r')\frac{R_y}{R^3} - J_y(r')\frac{R_x}{R^3}dv \tag{3.12af}$$

Using the equations above, let's simplify the equation for the magnetic flux density.

$$B = \frac{\mu}{4\pi}\int_v \nabla \times \frac{J(r')}{|r-r'|}dv \tag{3.12ag}$$

$$B = \frac{\mu}{4\pi}\int_v \frac{J(r') \times R}{|r-r'|^3}dv \tag{3.12ah}$$

If the cross-sectional area of the conductor stays constant, the $J\left(r'\right) = Idl$. Hence the equation above comes to its final form.

$$B = \frac{\mu}{4\pi} \int_l \frac{Idl \times R}{|R|^3}$$ (3.12ai)

Application 31: Finding the Magnetic Field Generated by a Loop Current Using the Biot-Savart Law

The Biot-Savart law is being applied heavily in determining the magnitude and direction of the magnetic field generated by ring or loop currents. This is especially an important law in loop antenna analysis.

Let's consider a ring current with a radius *ofa*, carrying a current of I in the a_ϕ direction as shown in Fig. 3.13. Let's consider the magnetic field at a height of h located on the z-axis.

Let's start with the position vectors of the observer and the source given in Eqs. 3.13a and 3.13b.

$$r = ha_z$$ (3.13a)

Fig. 3.13 Application of the Biot-Savart law for a loop current

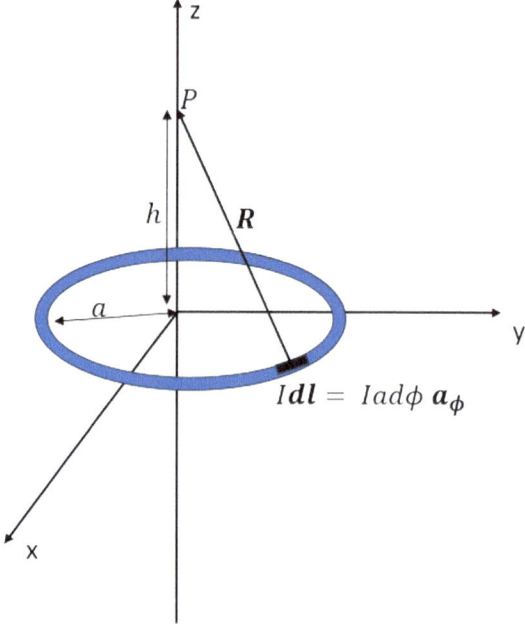

$$Idl = Iad\phi\ a_\phi$$

$$r' = aa_\rho \tag{3.13b}$$

$$R = -aa_\rho + ha_z \tag{3.13c}$$

$$dl = ad\phi a_\phi \tag{3.13d}$$

$$B = \frac{\mu}{4\pi} \int_l \frac{Idl \times R}{|R|^3} = \frac{\mu}{4\pi} \int_{\phi=0}^{2\pi} \frac{Iad\phi a_\phi \times (-aa_\rho + ha_z)}{\left(\sqrt{a^2 + h^2}\right)^3} \tag{3.13e}$$

$$B = \frac{\mu}{4\pi} \int_{\phi=0}^{2\pi} \frac{Ia^2 d\phi a_z}{\left(\sqrt{a^2 + h^2}\right)^3} + \int_{\phi=0}^{2\pi} \frac{Iahd\phi a_\rho}{\left(\sqrt{a^2 + h^2}\right)^3} \tag{3.13f}$$

The a_ρ directed component cancels out due to symmetry. The output becomes:

$$B = \frac{\mu}{4\pi} \frac{Ia^2 2\pi a_z}{\left(\sqrt{a^2 + h^2}\right)^3} = \frac{\mu Ia^2 a_z}{2\left(\sqrt{a^2 + h^2}\right)^3} \tag{3.13g}$$

If the medium of the loop is in air the equation becomes:

$$B = \frac{\mu_0 Ia^2 a_z}{2\left(\sqrt{a^2 + h^2}\right)^3} \tag{3.13h}$$

Application 32: Radio Frequency Identification—RFID

Radiofrequency identification works based on electromagnetic induction. This is also known as near-field coupling. RFID sensors or tags can be categorized as active and passive. Active RFID sensors have an internal power source to power them whereas passive sensors energize by electromagnetic induction. As shown in the Fig. 3.14 the RFID reader emits an alternating voltage signal towards the RFID sensor. Given the proximity of the reader and the sensor, there is an induced voltage on the solenoid of the RFID sensor. Inside the sensor, there is a load, controller, a clock, and a rectifier. As the voltage is being induced on the solenoid of the sensor it is rectified at the same time it activates an internal clock. The clock signal is used to switch on and off the power to the load connected to the sensor. This is known as load modulation. This modulated signal is then coupled back to the reader solenoid (antenna). Hence the reader recognizes the tag. The RFID tags used in stores use this near-field inductive coupling technique.

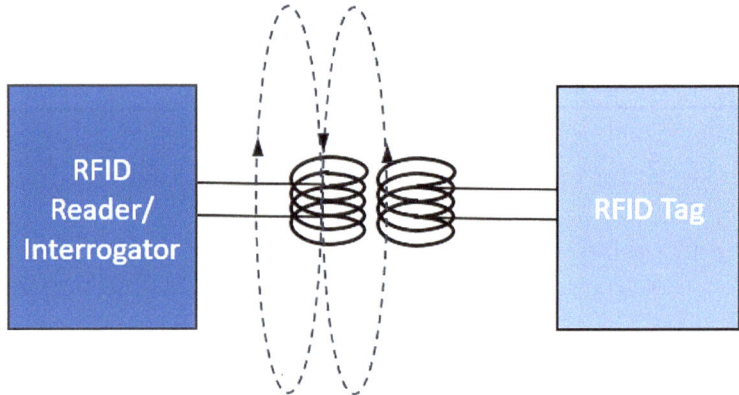

Fig. 3.14 The radio frequency identification system

References

- Website: https://www.mdpi.com/1424-8220/19/18/4012 accessed June 25, 2024.
- Website: https://www.camcode.com/blog/what-are-rfid-tags/#:~:text=RFID%20tags%
  20are%20small%20electronic,monitoring%20the%20movements%20of%20livestock.
  Accessed June 25, 2024.

Application 33: Magnetic Card Readers

Magnetic cards work using the principle of Faraday's Law. In a magnetic strip, the tiny
iron particles are arranged as small bar magnets with like poles adjacent to each other as
shown in Fig. 3.15. Faraday's Law states when there is a time-varying magnetic field it
induces a voltage in a closed loop. In this case, the closed loop is a solenoid connected
to a voltmeter. The time-varying behavior is provided by moving the magnetic strip in
the vicinity of the solenoid. When the magnetic strip is swiped the north poles induce a
voltage in one direction while the south poles induce the voltage in the opposite direction.

This voltage pattern is then converted to a discrete signal and then encoded into a
string of ones and zeros. The coding technique used in magnetic cards is known as F2F
coding scheme.

References

- Website: https://www.edn.com/design-a-cost-effective-magnetic-card-reader/ accessed
  June 25, 2024.
- Website: https://mallikarjuntirlapur.github.io/MagneticStripeReader/, accessed June 25,
  2024.

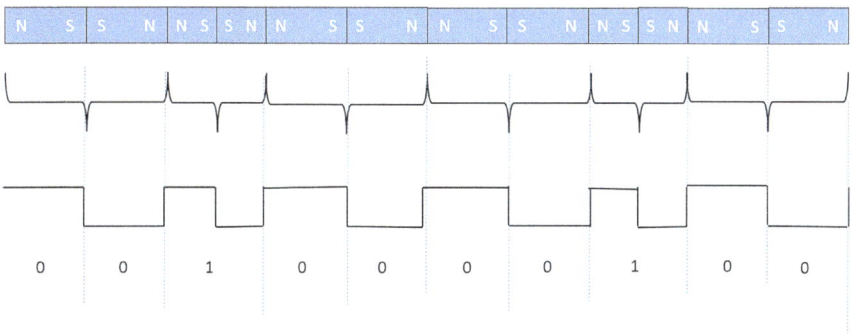

Fig. 3.15 Signals generated in magnetic card reading

Practice problems

1. A levitating globe uses an electromagnet to suspend a metal globe in the air.
    a. The bottom disk of the device creates an upward force which makes the metal globe elevated from its base. Assume that the above globe with a mass of 50 g was elevated 7 cm from the base with an acceleration of 0.1 m/s^2.
    b. Calculate the upward force exerted by the device at the elevating state.
    c. If the area of the electromagnet is 5 cm^2, how much of a magnetic field intensity can be observed in the air?
    d. Hence calculate the magnetic flux density measured in the air.
    e. The above magnetic flux density is produced by a solenoid wrapped around an iron core with a relative permeability of 5000. How much of a magnetic field intensity can be measured inside the core?
    f. Using the area of the electromagnet given in part b, calculate the total magnetic flux produced by the iron core.
2. An electric bell uses an electromagnet to produce sound. Once the switch is activated current starts flowing in the circuit, hence the electromagnet starts storing energy. The force produced by the electromagnet attracts the metal piece attached to the hammer. That force makes the hammer move and hit the gong to produce sound.

    Consider the cross-section of the electromagnet to be 5 cm^2 and the mean path length as 8 cm. an electromagnet is created by winding 12 turns of a wire around an iron core with a relative permeability of 5000. The current through the circuit is 10 mA.
    a. Calculate the magnetic flux inside the electromagnet.
    b. Hence calculate the magnitude of magnetic flux density measured inside the core.
    c. If the above flux is leaked into the air how much is the magnitude of magnetic field intensity in the air?

    d. How much of a force is exerted by the electromagnet on the hammer? (Effective area for force would be twice the cross-sectional area due to the geometry). Also moving of the hammer occurs in air medium.

3. Transformer EMF

    A typical transformer has two windings, the primary and the secondary. In the above transformer, there are 200 windings in the primary and 400 turns in the secondary. And the current through the primary winding is 10mA. If the core of the transformer is made with 99% pure iron with a relative permeability of 5000. The core has a square cross-section of 10 cm × 10 cm. The mean path length is 20cm.

    a. First, using only the primary winding, calculate the reluctance of the circuit.

    b. How much is the magnetic flux inside the core created by the primary winding?

    c. The magnetic flux through the secondary is the same as the flux through the core. Hence find the current through the secondary winding.

    d. How much is the reluctance from the secondary winding?

    e. Comment on the current ratio and the turn ratio between the primary and secondary windings.

4. Plasma globes are used as decorative elements. Inside the globe, there is a Tesla coil to produce a high voltage and trapped electrons to generate the plasma.

    a. For now, let's assume that the Tesla coil contains only one loop. If the voltage inside the plasma globe is 3000 V (real value) and if this voltage is induced along a circular Tesla loop with a diameter of 3 cm, how much is the electric field intensity?

    b. If the velocity of the electrons inside the plasma globe is 10 ms^{-1}, how much is the magnitude of the induced magnetic field?

    c. If there are 10 turns within the Tesla coil each has a diameter of 3 cm, and each produces an electric field intensity that you calculated in part a, how much would be the new total voltage produced by the coil?

    d. Hence how much is the new induced magnetic field?

5. Maglev, is a high-speed transportation mechanism. It uses magnetic circuits (for example toroid) to levitate the locomotive (for example train). Since there is almost zero friction, these locomotives can be driven at very high speeds.

    a. Consider the mass of the locomotive with people inside to be 8×10^7 kg. If the gravitational acceleration g is 10 m/s^2. How much is the weight of the locomotive?

    b. Magnetic circuitry should produce a force equal to the weight of the train to levitate it. Lifting occurs in the air. If the active area of the magnetic circuit and the locomotive is 100 m^2 how much is the magnetic field intensity produced by each magnetic circuit? Assume the total force is evenly distributed between the two circuits.

    c. If the lifting gap (air gap in the above diagram) is 10 cm, how much of magnetic energy is produced by one circuit?

    d. The magnetic field intensity in the air that you calculated in part b, is produced by a 99% pure iron core with relative permeability of 5000. Hence calculate the magnetic field intensity inside the core.

Eddy Current Applications

4

Eddy currents are linked to the Faraday's Law. Although eddy currents are a by-product they have a wide range of applications. This chapter discusses the applications of eddy currents.

Application 34: Skin Effect and Eddy Currents

Faraday's law says when the magnetic field across a loop changes, it induces a voltage. This voltage is called the electromotive force since it is the force needed to mobilize a unit charge. And the EMF generates a current through the conductive loop as we studied above in the generators. This current in turn generates a magnetic field in the opposite direction of the original magnetic field, hence the negative sign in Faraday's Law, which was introduced by Lenz. What if there is a metallic surface instead of a loop? In that case, still, there is an induced electromotive force. And this electromotive force will produce a current that looks like a loop. These circular looped currents are called eddy currents since they look like eddies in water. Similar to the induced currents in the motional electromotive force, the eddy currents generate a magnetic field in the opposite direction to the original magnetic field.

Consider the Fig. 4.1 where a conductor is moving into the field of a permanent magnet. The left side of the conductor is entering the magnet. Hence the magnetic field on the left side of the conductor is increasing. To oppose that increase there will be a counterclockwise current on the left side of the conductor. When you point the thumb of your right hand in the direction of the current, the fingers curl in the direction of the magnetic field. The right side of the conductor is leaving the magnetic field; hence the magnetic

© The Author(s), under exclusive license to Springer Nature Switzerland AG 2025
A. Maxworth, *One Hundred Applications of Maxwell's Equations*, Synthesis Lectures on Electromagnetics, https://doi.org/10.1007/978-3-031-73784-8_4

Fig. 4.1 Directions of eddy currents on a metallic plate moving under a magnet

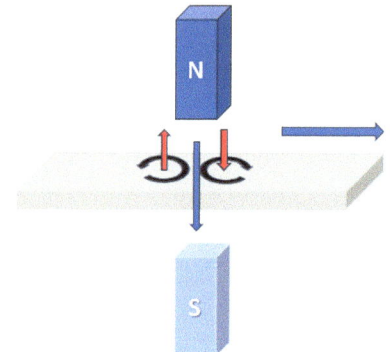

field is decreasing. On the right side of the conductor, there will be an eddy current in the clockwise direction, and they produce a magnetic field in the same direction as the original magnetic field.

The distribution of eddy currents within the skin depth of a conductor is known as the skin effect. These eddy currents create heat within the skin of the conductor increasing its effective surface resistance. This makes energy loss known as the eddy current loss. One way to minimize these eddy current losses is by reducing the surface area of the conductor interacting with the magnetic field.

The eddy currents can be desirable or undesirable. When it comes to transmission lines and antennas, eddy current losses are undesirable since they reduce the transmitted energy.

References

- Website: https://courses.lumenlearning.com/suny-physics/chapter/23-4-eddy-currents-and-magnetic-damping/ accessed June 25, 2024.
- Website: https://farside.ph.utexas.edu/teaching/316/lectures/node89.html accessed June 25, 2024.

Application 35: Non-destructive Testing or the Eddy Current Testing

Nondestructive testing (NDT) uses the same principle as metal detectors. The metal detector detects the drop in the induced magnetic field in the detector coil. Similarly in non-destructive testing, the change in the magnetic field is detected. In the presence of a time-varying magnetic field, there is an induced voltage called the electromotive force

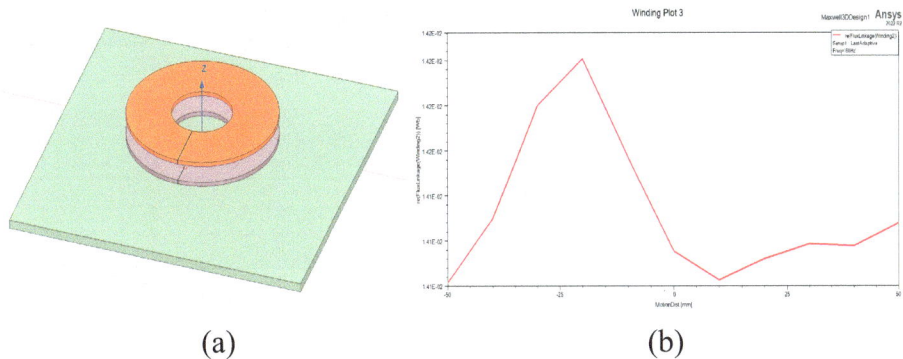

(a)   (b)

Fig. 4.2 **a** The nondestructive testing coils. The top coil is the transmission coil, and the middle coil is the detector coil. The metal plate has an opening simulating the damage. **b** Shows the flux linkage graph of the detector coil. As the damaged area of the metal plate enters the detector the flux linkage increases and as the damaged area passes through the detector the flux linkage reduces

on the conductive material. This electromotive force can create circular currents on metallic surfaces. The direction of this current is such that, the magnetic field it generates will be in the opposite direction to the original magnetic field.

In non-destructive testing, the detector coil measures the magnetic flux through it in the same way as the search coil of the metal detector. The difference is, in non-destructive testing, there will always be eddy currents since there are metallic surfaces. In the presence of structural damage such as a crack, the eddy current produced by the metallic sheet will be different, hence there will be a change in the magnetic flux produced by the cracked portion of the metal sheet, changing the total magnetic field through the detector coil.

Non-destructive testing can be used on multi-layer material as well. The drawback is it can only detect damages perpendicular to the surface. If damage is parallel to the surface, it will not be detected. Figure 4.2a shows a model of a non-destructive eddy current testing system created using Ansys Maxwell 3D. Figure 4.2b show the magnetic flux linkage through the detector coil which is the second coil from the top. As seen from the graph, as the damaged area is passing through the detector the flux linkage increases due to the lack of eddy currents.

Application 36: Wireless Power Transfer or the Wireless Charger

The wireless charger works using the same principle as the transformer. In a wireless charger, there are two coils: one is in the power source which we may call the primary and another is in the equipment that needs to be charged which we call here the secondary. The primary coil is connected to an alternating current/voltage. This time-varying current generates a time-varying magnetic field.

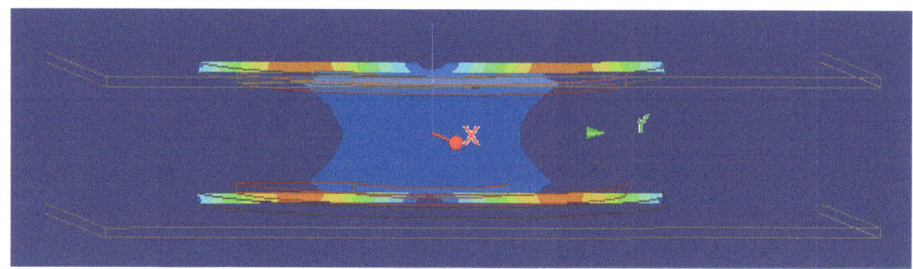

Fig. 4.3 Magnetic flux linkage simulation using Ansys Maxwell 3D. The Bottom coil is the primary coil, and the top coil is the secondary

The equipment that needs to be charged such as a cell phone is kept in the vicinity of this time-varying magnetic field, it induces a voltage in the secondary coil. The Fig. 4.3 shows the magnetic flux linkage from the primary coil (bottom) to the secondary (top) modeled in Ansys Maxwell 3D. The working principle of the wireless charger is the same as a step-down transformer.

Application 37: Metal Detector

The metal detector works based on transformer EMF. Here there are two coils—a primary coil which generates a time-varying magnetic field and a detector coil similar to the secondary which will have an induced voltage (EMF) because of the magnetic flux linkage. The detector coil does not have any direct power connection (as in a transformer) and it detects the induced magnetic field from the primary or the source coil.

When there are metal components, an electromotive force is induced on those components as well. This induced electromotive force generates currents within these metal components. If there are metal loops present the induced EMF generates an induced current which in turn generates a magnetic field in the opposite direction of the original magnetic field. On metallic surfaces, the induced electromotive force generates eddy current loops. Eddy current loops are also formed in a way that the magnetic field generated due to eddy currents is opposite to the original magnetic field from the source coil. Due to these magnetic fields induced on metallic components, the magnetic flux through the detector coil drops. When there is a drop in the detected magnetic flux through the detector coil, due to the induced magnetic field in the opposite direction of the source coil, the detector emits a beep sound to indicate the presence of a metal.

Figure 4.4a shows a simple metal detector with the transmission (top) and detector (bottom) coils. When the metallic slab is going through the detector, the flux through the coil drops.

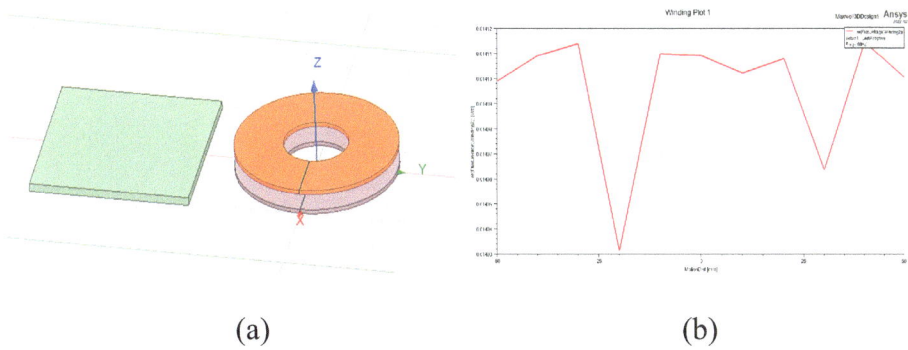

(a)   (b)

Fig. 4.4 **a** A metal detector with transmission and detector coils and a metal slab is passing through the detector. The simulation was done in Ansys Maxwell 3D. The graph in **b** shows the reduction in magnetic flux linkage through the detector coil due to the presence of metal

In airport metal detectors, the areas where there are drops in magnetic fields are marked on a diagram.

Reference

- Website: https://www.brainkart.com/article/Application-of-eddy-currents_38497/ accessed June 25, 2024.

Application 38: Induction Cooking

Induction cooking uses the heat generated from eddy currents for cooking. In induction cooking, there is a coil on the source side made with a highly conductive material such as copper or aluminum. The purpose of the coil is to conduct current and generate a time-varying magnetic field. Since the coil stays stationary, the source should be an alternating current source. This time-varying magnetic field induces an electromotive force on the surface of the vessel that needs to be heated. The flux linkage process is the same way as a transformer here, consider the bottom of the vessel as a coil of one turn. Given the transformer equation $N_p i_p = N_s i_s$, when the number of turns in the secondary winding is one, the current in the secondary increases. Therefore, there is a very high induced current at the bottom of the cooking vessel. This induced current is in the form of a loop or an eddy at the bottom of the cooking vessel. The direction of this eddy current is in a form such that it creates a magnetic field in the opposite direction of the original magnetic field of the cooker. The Fig. 4.5 shows the direction of the magnetic field from the induction cooker and the direction of the eddy currents modeled using Ansys Maxwell 3D. Notice that when you used the right-hand rule pointing the thumb of the right hand

Fig. 4.5 The direction of the magnetic field from the electromagnet and the eddy currents induced on a metallic slab modeled using Ansys Maxwell 3D

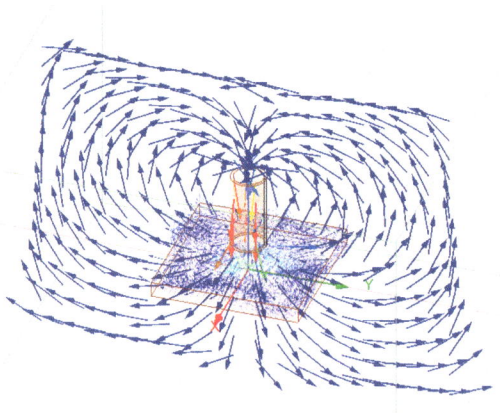

in the direction of the current, the magnetic field generated by the eddy current is in the opposite direction of the original magnetic field.

The vessels used in induction heating are made with a ferromagnetic material such as cast iron. The ferromagnetic material has high magnetic permeability making it able to generate internal magnetic field lines and concentrate the heat into the skin depth of the vessel. Eddy current heating is more power efficient compared to conduction heating.

Application 39: Electromagnetic Brake

Electromagnetic breaks are being used to smoothly slow down large vehicles such as trains. The electromagnetic brake designs can vary. In some electromagnetic brakes, a rotating disk or a break plate is rotating through a permanent magnet. As the rotor rotates through the permanent magnet, the eddy currents are generated, and these eddy currents create a magnetic field opposite to the direction of the magnetic field. The force generated from these eddy currents is in the opposite direction to the direction of rotation—slowing the rotation.

Another type of electromagnetic brake is the electro-magnetic failsafe brake which is used in aircraft doors and medical instruments. The Fig. 4.6 shows an electromagnetic failsafe brake with a steel core, a copper coil which is an electromagnet, and the brake plate. There is a shaft that goes through the center of the brake, but it is not shown in the design since it is not magnetic. When the electromagnet is energized, it attracts the brake plate releasing the shaft, hence releasing the door. As the electromagnet is activated there are eddy currents generated on the brake plate. These eddy currents in turn generate a magnetic field in the opposite direction of the initial magnetic field. The eddy current generated magnetic field dampens the attraction force—slowing the opening of the aircraft door and avoiding a sudden opening.

Fig. 4.6 An electromagnetic
failsafe brake modeled using
Ansys Maxwell 3D

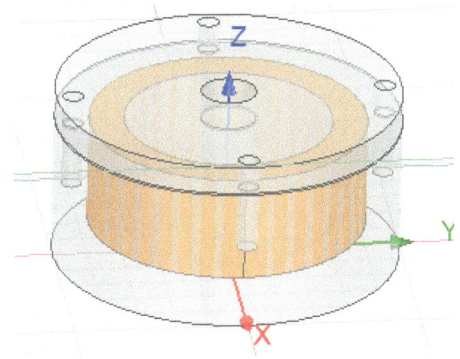

Application 40: Eddy's Current Damper

The purpose of the eddy current damper is to reduce the vibration of the structure. These dampers are used in metallic beams and scales for balance. The Fig. 4.7 shows a metallic beam. At the edge of the beam, there is a copper strip. The free end of the beam is free to move within a magnetic field created by a permanent magnet. When the beam vibrates within the magnetic field, the eddy currents are generated on the metal strip—producing a magnetic field in the opposite direction to the magnetic field from the permanent magnet. This induced magnetic field dampens the net magnetic field, dampening the vibrations.

In newer designs, instead of the passive copper strip is replaced by a copper coil connected to an alternative power supply making the coil active. In that scenario, by adjusting the current through the coil the strength of the magnetic field can be changed increasing the efficiency of the damper.

Application 41: Eddy's Current Separator

Eddy current separators are used in a wide range of applications including detecting counterfeit coins, sorting aluminum cans from plastic bottles at the recycling centers, sorting metallic components from garbage, etc.

The Fig. 4.8 an eddy current separator, separating ferromagnetic materials, aluminum cans, and other non-metallic objects. At the end of the conveyor belt, there is a rotating magnet. When the magnet rotates, eddy currents are induced on the aluminum (or other conducting material) cans. The magnetic field generated by the eddy currents is in the opposite direction of the original magnetic field. In other words, the eddy current generated magnetic field repels the original magnetic field. Hence, the aluminum cans are repelled away from the conveyor belt. The non-metallic material will be dropped to the closer bin due to gravity. The ferromagnetic material with high permeability will be attracted to the magnet through the thin conveyor belt, and those should be scraped off.

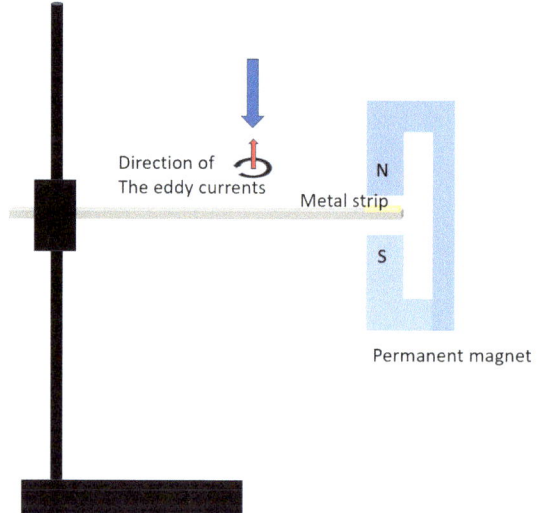

Fig. 4.7 The eddy current damper

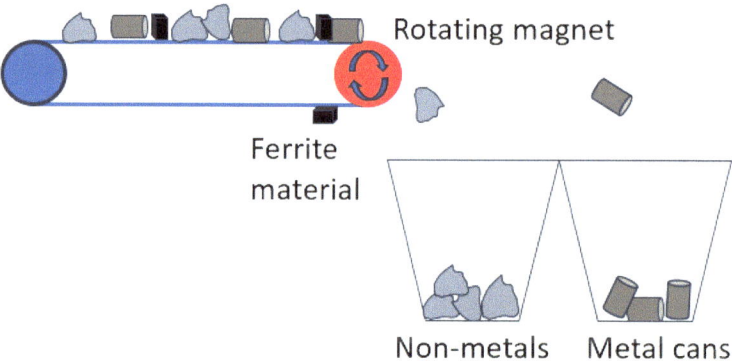

Fig. 4.8 Eddy current separator

Application 42: Coating Thickness Measurement

Eddy currents are one way of measuring the coating thickness. The coil setup in this case is the same as the metal detector. A primary coil with a high-frequency alternating current is brought close to the coated non-ferrite metallic surface. As expected, these alternating currents generated eddy currents on the coated metallic surface. The eddy currents generate a magnetic field in the opposite direction of the original magnetic field (the source magnetic field). This results in the dampening of the total magnetic field sensed by a secondary coil.

With the increasing coating thickness, the eddy currents generated on the metallic surface reduce, reducing the secondary magnetic field generated by the eddy currents. Hence the net magnetic field measured by the secondary coil will be close to the original magnetic field with less attenuation.

If the coating thickness is small, then the generated eddy currents can be strong, which in turn creates a strong secondary magnetic field weakening the net magnetic field measured by the sensing (secondary) coil. This method is known as amplitude-sensitive eddy current testing.

Before using the eddy current-based coating thickness measurements, it is important to calibrate the instruments. The equations used in thickness calculations differ based on the geometry of the structure and the material properties. Therefore, depending on the situation there are different standards to be used as given in the references.

References

- Website: https://www.defelsko.com/resources/coating-thickness-measurement#:~:text= Eddy%20current%20techniques%20are%20used,surface%20of%20the%20instrument 's%20probe accessed June 25, 2024.
- Website: https://www.helmut-fischer.com/en/applications/solutions/amplitude-sensit ive-eddy-current-method?setLang=en&cHash=a4c7382a54685f43bfc51ec908a2a4e1 accessed June 25, 2024.
- Website: https://www.nde-ed.org/NDETechniques/EddyCurrent/AdvancedTechniques/ remotefieldsensing.xhtml accessed June 25, 2024
- Website: https://www.nde-ed.org/NDETechniques/EddyCurrent/AdvancedTechniques/ scanning.xhtml accessed, June 25, 2024

Application 43: Electrodynamic Levitation—Electrodynamic Bearing

Electrodynamic bearings are being used in applications such as electricity meters to wind turbines. The Fig. 4.9 shows an electrodynamic bearing that is used in large-scale applications. The bearing shown in the figure has two types of bearings: radial bearings and axial bearings. The radial bearings keep the rotating shaft levitated at the center of the bearing and the axial bearings keep the conductive plates connected to the rotating shaft levitated at the center of the bearings. The electrodynamic bearings can be active or passive. The passive bearings use permanent magnets, and the active bearings use electromagnets in which the magnetic field can be controlled actively using electric current. The bearing system shown in the figure uses passive axial bearings and active radial bearings. In magnetism, the like polarities repel each other producing the levitation. In the case of electrodynamic bearings, we use the magnetic field produced by the eddy currents to produce the levitation.

Remember, from Faraday's law, when there is a rate of change of magnetic flux in a certain direction, the induced electromotive force is being produced. The current generated as a result of this electromotive force produces a magnetic field to oppose change of the original magnetic field (Lenz's Law). In this case, there will be loops of currents known as the eddy currents on the rotating shaft and the conductive plates connected to the rotating shaft. These eddy currents will generate magnetic fields to oppose the change of the magnetic flux from the source magnets. The axial levitation is like the magnetic damper.

References

- Website: http://www.lim.polito.it/research/electrodynamic_bearings, accessed June 25, 2024.
- Website: https://www.youtube.com/watch?v=hJ6fXGjz4Ik accessed June 25, 2024.
- Website: https://www.youtube.com/watch?v=IAkxS1xVraw accessed June 25, 2024.

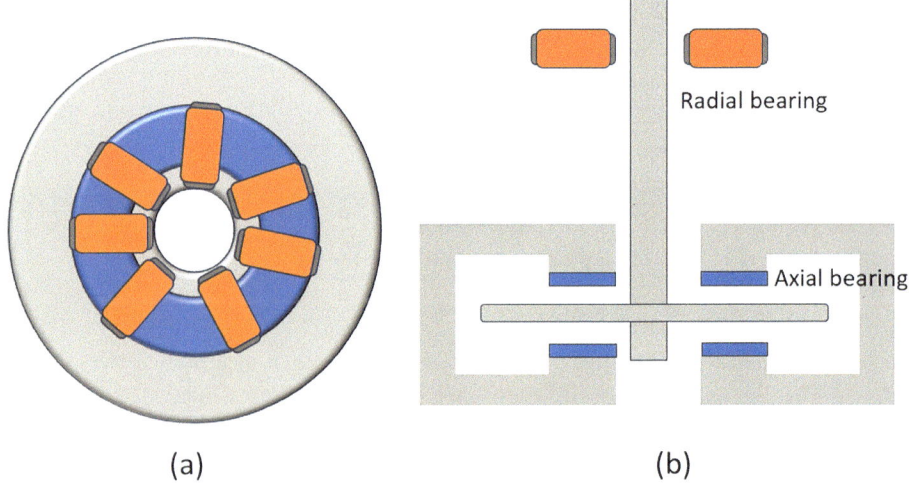

(a)   (b)

Fig. 4.9 **a** The top-side view of electrodynamic bearing. **b** The cross section of electrodynamic bearing

Practice problems.

1. Induction cooking.
   a. The AC source operates an induction cooker at a frequency of 10 kHz, and it generates a magnetic flux density of 25 μT. Calculate the magnitude of the rate of change of magnetic flux density.
   b. The coil here has 9 turns and a radius of 5 cm. Calculate the rate of change of magnetic flux. Remember to multiply by the number of turns.
   c. The bottom of the frying pan has a radius of 8 cm. Calculate the electric field intensity at the base of the frying pan.
   d. If the switching frequency of the AC source was increased to 20kHz while keeping all the other parameters the same, how much is the new EMF?
2. Eddy current braking system or Faraday braking system uses Faraday's Law of Induction and Ampere's Law on Magneto Motive Force (MMF).

   In this braking system, there is a rotating disk in between an electromagnet. Eddy currents are closed-loop currents on the surface of the rotating body perpendicular to the magnetic field from the electromagnet. You do not need to know about eddy currents for this problem, other than the fact that eddy currents generate an opposing magnetic field to the field generated by the electromagnet.

   The magnetic force is provided by the electromagnet and the eddy currents create the antimagnetic force. $\mathfrak{R}_c, \mathfrak{R}_a,$ and $\mathfrak{R}_d,$ are the reluctances of the electromagnet core, air, and the rotating disk respectively. Φ_c is the magnetic flux that goes through the entire circuit. The electromagnet has a ferrite core, and the rotating disk is made of copper.
   a. If the electromagnet has 3350 windings, and the current through it is 3A, using Ampere's law calculate the magneto-motive force MMF from the electromagnet.
   b. Out of the three reluctances, $\mathfrak{R}_c,$ is the lowest and hence neglected in calculations. If the gap between the two magnetic poles (l_g) is 7 mm and the thickness of the copper disk (d) is 4mm, calculate the reluctances of air $(\mathfrak{R}_a,)$ and the disk $(\mathfrak{R}_d,)$. The relative permeability of copper is 1. The cross-sectional area of the magnetic flux (A) is 1600 mm^2.
   c. The magnetic flux for the eddy current braking system is given by:

$$\Phi_c = \frac{MMF\,from\,the\,electromagnet}{\mathfrak{R}_a, + \mathfrak{R}_d, + \sigma.a.d.r.\omega/A}$$

σ = conductivity of copper at 200 °C = 32 Mega Siemens/meter

a = distance between the center of the electromagnet to the center of the disk = 150 mm

d = thickness of the disk = 4 mm

r = radius of the disk = 22.57 mm

ω = angular frequency of the disk = 6000 rad/s

A = cross-sectional area of the electromagnet or the cross-sectional area of the magnetic field = 1600 mm^2

Hence calculate the magnetic flux through the circuit Φ_c.

d. The anti-magnetomotive force from the eddy currents is given by:

$$anti\,MMF \;=\; MMF\,from\,the\,electromagnet \;-\; \Phi_c(\mathfrak{R}_{a,} + \mathfrak{R}_{d,})$$

Hence calculate the anti-magnetomotive force from the eddy currents.

e. Calculate the magnetic flux density through the disk.

Lorentz Force

5

Lorentz force is not a direct derivation of Maxwell's equations although it could be derived using Maxwell's equations. The applications of the Lorentz force and the Maxwell's equations are linked. This chapter discusses some of the applications of the Lorentz force.

Application 44: Lorentz Force

The Lorentz force is named after the Dutch physicist Hendrik Lorentz (1853–1928). Lorentz's force expresses the electromagnetic force applied on a charged particle by the electric and magnetic fields. The Lorentz force originates from James Clerk Maxwell's and Oliver Heaviside's contributions to the electric and magnetic forces. Since Hendrik Lorentz did the full derivation of this formula, it was named in his honor.

The Lorentz force has applications ranging from electromagnetic propulsion to plasma physics. The Lorentz force is defined as:

$$F = q(E + v \times B) \tag{5.1}$$

In the above equation, q is the electric charge, E is the electric field intensity, v is the velocity of the charged particle, and B is the magnetic flux density. And F is the Lorentz force. If there are no magnetic fields and only an electric field exists, the Lorentz force equation reduces to:

$$F = qE \tag{5.1a}$$

Which is the Coulomb's law for electric charges. The Columb's force is defined as:

© The Author(s), under exclusive license to Springer Nature Switzerland AG 2025
A. Maxworth, *One Hundred Applications of Maxwell's Equations*, Synthesis Lectures on Electromagnetics, https://doi.org/10.1007/978-3-031-73784-8_5

Fig. 5.1 A conducting
armature moving inside a
magnetic field

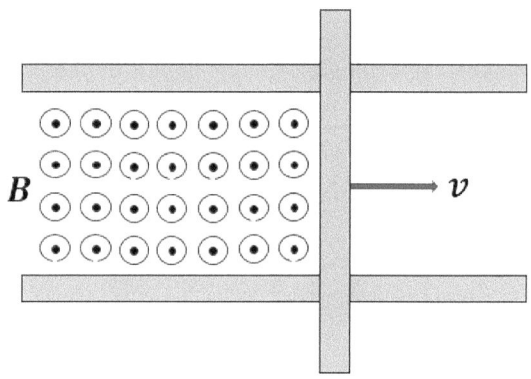

$$F = \frac{q_1 q_2}{4\pi \varepsilon_o r^2} a_r \tag{5.1b}$$

The electric field intensity is the Coulomb's force acting on a unit test charge. Hence, the force acting on a q charge is the multiplication of the electric field intensity by the charge gives the total Columb's force acting on it.

Now, let's consider the relationship between the Lorentz force and Faraday's Law. Let's consider the setup shown in Fig. 5.1. Where there are two conductive rails and a moving conducting armature. The armature moves to the a_x, direction with a velocity of v. The direction of the magnetic field inside the loop created by the conductive rails and the armature is shown in the diagram.

Let's find the electromotive force induced across the conductive by the motion of the armature. The Faraday's law says:

$$\oint_l E.dl = -\frac{\partial \phi_m}{\partial t} \tag{5.1c}$$

$$E_x \Delta x a_x + E_y d a_y - E_x \Delta x a_x = -\frac{\partial (\Delta x dB)}{\partial t} \tag{5.1d}$$

$\Delta x dB$ indicates the total magnetic flux within the loop. Let's get everything in vector form.

$$E_y d a_y = -\frac{\Delta x}{\Delta t} a_x \times B a_z d \tag{5.1e}$$

$$E_y a_y = -v a_x \times B a_z \tag{5.1f}$$

$$E_y a_y = v B a_y \tag{5.1g}$$

In other words, Faraday's Law says that when a conductor is moved within a magnetic field such that there is a rate of change of magnetic flux, there will be an induced potential or the motional electromotive force. The magnitude of that electric field intensity would be vB and the direction of that electric field intensity would be perpendicular to both the velocity and the magnetic flux density.

Now, let's use the Lorentz force and derive the electric potential induced on the moving armature. When the armature is moving at a constant velocity, the force acting on the armature is zero. Hence:

$$F = q(E + v \times B) = 0 \tag{5.1h}$$

$$E = -v \times B \tag{5.1i}$$

$$E_y a_y = -v a_x \times B a_z \tag{5.1j}$$

$$E_y a_y = v B a_y \tag{5.1k}$$

The electric potential of the moving armature is:

$$-\int_{y=d}^{0} E_y . dy = \Phi = vBd \tag{5.1l}$$

Lorentz Magnetic Force and the Average Magnetic Force

The Lorentz force is the net electromagnetic force generated by the electric and magnetic fields on a charge of q. As we say earlier the electric force in the absence of magnetic fields is the same as the Colomb's force.

$$F_E = qE \tag{5.1m}$$

Now, let's compare the magnetic forces from the Lorentz force equation and the magnetic force derived through the average magnetic energy equation.

$$W_{magnetic} = \frac{1}{2} \int_v H.B dv = \frac{1}{2} \int_v H.B dA.dl \tag{5.1n}$$

$$W_{magnetic} = F_m.l \tag{5.1o}$$

And the magnetic force becomes:

$$F_m = \frac{1}{2} \int_A H.BdA \tag{5.1p}$$

The $\frac{1}{2}$ is because is to avoid double counting the force from one pole to another. Hence we can ignore it for now.

Let's consider the area element as the vector product of two length elements.

$$F_m = \int_{l_2} \int_{l_1} H.Bdl_1 dl_2 \tag{5.1q}$$

$$F_m = \int_{l_2} H.dl_2 \int_{l_1} B.dl_1 \tag{5.1r}$$

From the Ampere's Law $.\int_{l_2} H.dl_2 = I_{enclosed}$

And the current is the rate of change of charge $\frac{dq}{dt}$. Also, remember that the magnetic field and the current are perpendicular to each other. Assuming that all variables are independent of the length l_1, let's rearrange the above equation.

$$F_m = \int_{l_1} dl_1 I \times B \tag{5.1s}$$

$$F_m = \int_{l_1} dl_1 \frac{dq}{dt} \times B = q \int_{l_1} \frac{dl_1}{dt} \times B = qv \times B \tag{5.1t}$$

The magnetic force expression derived for the magnetic force from the average magnetic work, and the magnetic work from the Lorentz force are the same.

Application 45: Magnetic Flow Meter—Mag Meter

The magnetic flow meter shown in the Fig. 5.2 is used in measuring the volumetric flow rate of conductive fluids such as seawater, wastewater, and blood. The magnetic flow meter works based on the Lorentz force.

Let's consider a flow tube with a cross-section area of A. The conductive fluid passing through the tube has both positive and negatively charged ions. There is a magnetic field acting downwards through the conductive fluid. There is a pair of electrodes perpendicular to both the magnetic field and the direction of the flow. Let's consider the distance between the electrodes as d.

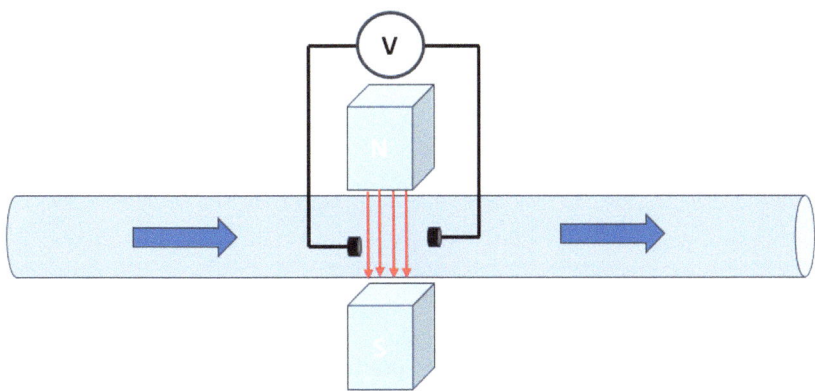

Fig. 5.2 A schematic of a magnetic flow meter

Let's consider the Lorentz force acting on both positive and negatively charged ions as they pass through the magnetic field. The magnetic force acting on the charges is:

$$\boldsymbol{F_m} = q\boldsymbol{v} \times \boldsymbol{B} \tag{5.2}$$

According to the above equation, as the conductive fluid passes through the magnetic field, the positive ions experience a force towards the back of the tube while the negative ions and electrons experience a force towards the front of the tube. This separation of charges creates an electric field intensity measured by the two electrodes. This is an induced electric field due to the charge movement through the magnetic field. This induced field can be deduced from the Lorentz force as well.

$$\boldsymbol{E} = -\boldsymbol{v} \times \boldsymbol{B} \tag{5.2a}$$

This electric field intensity is between the two electrodes within a separation of d. The potential difference between the two electrodes is:

$$\Phi = \boldsymbol{v} \times \boldsymbol{B} \times d \tag{5.2b}$$

In a flow meter system, the magnetic flux density, the distance between the electrodes and the cross-sectional area of the tube are known parameters. Therefore, by measuring the induced voltage across the electrodes the flow velocity can be calculated.

$$|\boldsymbol{v}| = \frac{\Phi}{|\boldsymbol{B}|d} \tag{5.2c}$$

The volumetric flow rate can be calculated by multiplying the velocity of the fluid by the cross-sectional area.

Reference

- Website: https://www.youtube.com/watch?v=D999KDUj_QU accessed June 25, 2024.

Application 46: Hall Effect Sensor

The Hall effect sensor works based on the principle of Lorentz's force. A Hall sensor is comprised of a semiconductor, a DC supply, and a voltmeter. Before moving into the functionality of the Hall sensor, let's consider the characteristics of semiconductors.

The semiconductors are made of group IV elements of the periodic table such as Carbon, Silicon, etc. The group IV elements of the periodic table have four electrons in their valance band. Hence these elements share four electrons from neighboring four atoms to achieve stability. When an electric field is applied these electrons become mobile creating a current. When an electron leaves the void created at that location is known as a hole. A pure semiconductor (intrinsic semiconductor) has an equal number of electrons and holes.

To increase the number of charge carriers in a semiconductor, "impurities" are added to it. Adding impurities is called doping. These impurities are elements from either group III or group V of the periodic table. If an intrinsic semiconductor is doped with a group III element, the resulting extrinsic semiconductor has more holes or positivity compared to an intrinsic semiconductor. Hence such an extrinsic semiconductor is called a p-type semiconductor. Similarly, if the intrinsic semiconductor is doped with a group V element, the resulting extrinsic semiconductor carries more electrons hence called n-type semiconductors.

In a Hall sensor, the DC supply makes the charges within the semiconductor flow in a certain direction creating a current flow. When there is no magnetic field present, the semiconductor connected to the DC supply acts as a typical wire.

But, when the semiconductor is exposed to a perpendicular magnetic field, according to the Lorentz force the charges experience a force perpendicular to both the magnetic field and the direction of the current (velocity). The direction of the force is different for both the electrons and the hole (positive charges). This Lorenz force induces a potential difference between the top and bottom edges of the semiconductor slab as shown in Fig. 5.3. This induced potential difference is called the Hall voltage. The generation of Hall voltage is called the Hall effect.

The Hall voltage changes as the strength of the magnetic field and the distance between the semiconductor and the magnetic field. Hence the Hall sensors are widely used as position sensors. Let's start with the magnetic force imposed on the charges on the semiconductor.

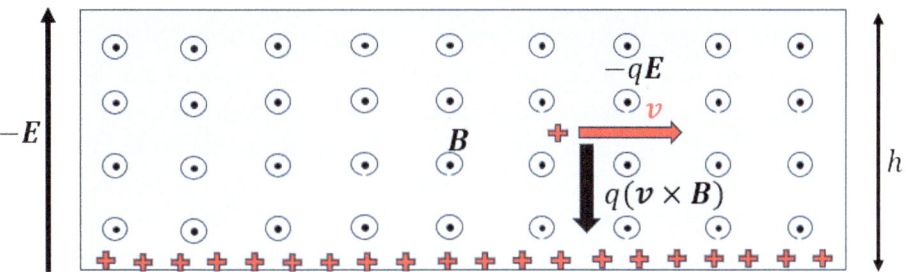

Fig. 5.3 The internal structure of a Hall-Effect sensor with a p-type semiconductor

$$F_m = qv \times B \qquad (5.3)$$

This magnetic field separates the electrons and holes and pushes them towards the opposite edges of the semiconductor. Once the electric potential builds up to balance the magnetic force the process stops. This can be explained by the Lorentz force.

$$F = q(E + v \times B) \qquad (5.3a)$$

when

$$E = -v \times B \qquad (5.3b)$$

$$qE = -qv \times B \qquad (5.3c)$$

$$F_E = F_m \qquad (5.3d)$$

and the potential difference is: $\Phi = vBh$

Figure 5.3 shows the directions of the magnetic force and the electric force for a p-type semiconductor.

References

- Website: https://pressbooks.online.ucf.edu/phy2053bc/chapter/the-hall-effect/ accessed June 25, 2024.
- Website: https://www.electronics-tutorials.ws/electromagnetism/hall-effect.html accessed June 2024.

Application 47: The Railgun

The railgun works based on the magnetic force. Since we explored the relationship between the Lorentz magnetic force and Faraday's Law of motional electromotive force we can also say it works based on Faraday's Law. The difference between the railgun and a generator is that a generator uses the motional electromotive force to generate a voltage (hence the name) and the railgun uses a voltage to generate a force. Let's express the magnetic force from the Lorentz force equation in terms of current, length, and magnetic flux density.

$$\boldsymbol{F_m} = q\boldsymbol{v} \times \boldsymbol{B} \tag{5.4}$$

$$\boldsymbol{F_m} = q\frac{d\boldsymbol{l}}{dt} \times \boldsymbol{B} \tag{5.4a}$$

$$\boldsymbol{F_m} = q\frac{d\boldsymbol{l}}{dt} \times \boldsymbol{B} = \frac{dq}{dt}d\boldsymbol{l} \tag{5.4b}$$

$$\boldsymbol{F_m} = I\boldsymbol{l} \times \boldsymbol{B} \tag{5.4c}$$

Note that the current is along the length of the armature, hence they are in the same direction, and the magnetic flux density is perpendicular to both. The following expression gives the magnitude of the magnetic force at any orientation of the current and the magnetic field, where θ is the angle between the current and the magnetic field.

$$|\boldsymbol{F_m}| = IlBsin\theta \tag{5.4d}$$

The Fig. 5.4 shows a railgun arrangement with conducting rails, a conducting armature (sabot), and the projectile. The current flow through the rails and the armature is shown in the figure. It also shows the magnetic field generated as a result of the current through the conductors. The force acting on the armature can be calculated using the above formulas originating from the Lorentz magnetic force. At the end of the conductive rail the armature breaks by hitting the aperture releasing the projectile.

As seen from the equation above, the Lorentz force is proportional to the length of the armature and the current through the circuit. The magnetic field is a current-dependent variable. Since there are limitations in increasing the length of the rail, the railguns use heavy currents (~1,000,000 A) to increase the Lorentz force.

Reference

- Website: https://science.howstuffworks.com/rail-gun1.htm accessed June 25, 2024.

Fig. 5.4 The railings and the conducting armature of a railgun

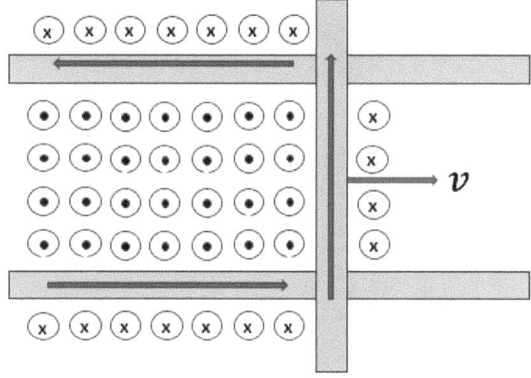

Application 48: Electrodynamic Propulsion—the Gaussian Coil Gun

The Gaussian coil gun is named after Carl Gauss who first studied the behavior of magnetic acceleration inside cannons. As the name implies, the coil gun uses a coil for initial acceleration and then a series of coils to accelerate the projectile. The projectile is made with a ferromagnetic material. Ferromagnetic materials have a high magnetic permeability. Hence those materials align with the magnetic field. When a ferromagnetic projectile is inside an air-filled electromagnetic solenoid, regardless of the direction of the magnetic field, it gets attracted to the center of the solenoid, since the center of the solenoid has the highest magnetic field strength.

When a projectile is released inside an electromagnetic solenoid it accelerates towards the center of the solenoid. Once it passes through the center of the solenoid, it starts decelerating. And accelerates toward the center by moving in the opposite direction. This is known as suck-back. This repetitive acceleration towards the center from either side, causes the projectile to oscillate around the center of the solenoid.

The coil gun working principle is, that as soon as the projectile reaches the center the power to the electromagnet, or the solenoid is shut off. Hence the projectile is released with a horizontal acceleration. As the projectile reaches the inlet of the second coil, it accelerates again toward the center and as soon as the projectile reaches the center the power to the second coil is turned off. This process can repeat for multiple stages.

Let's calculate the exit velocity of the projectile from this coil gun. First, we need to remember that the projectile is made of ferromagnetic material hence able to magnetize itself. The ratio between magnetization M and the magnetic field intensity H is given by the magnetic susceptibility χ_m. First, let's consider some of the basic relationships.

$$\mu = \mu_o(1 + \chi_m) \tag{5.5a}$$

$$\chi_m = \frac{M}{H} \tag{5.5b}$$

$$B = \mu H = \mu_o(H + M) \tag{5.5c}$$

The magnetic attraction force acting on this Ferromagnetic projectile is:

$$F_m = \mu_o V M \nabla(H) \tag{5.5d}$$

In Eq. 5.5d, V is the volume of the ferromagnetic projectile. The magnetic field intensity H, of a solenoid is:

$$|H| = \frac{NI}{h} \tag{5.5e}$$

In this case, let's define $n = \frac{N}{h}$, which is the number of turns of the solenoid per unit length. And the magnetic field gradient is $|H_{inside} - H_{outside}| = nI$. The attraction force acting on the ferromagnetic projectile is then converted into kinetic energy. Hence,

$$\frac{1}{2}mv_{exit}^2 = \mu_o V \chi_m n^2 I^2 \tag{5.5f}$$

$$v_{exit} = \sqrt{\frac{2\mu_o V \chi_m n^2 I^2}{m}} \tag{5.5g}$$

Reference

- Mathias Stolarski, 2011. "Die Magnetfeldüberlagerte Zentrifugation, Ein neues hybrides Trennverfahren zur Selektiven Bioseparatio"

Application 49: The DC Linear Induction Motor

The linear motor and the railgun work based on the same principles. Like the railgun, the linear motor also has two railings and a wire which acts as the armature of the motor. The difference is, in the railgun, the magnetic field is created by the current, and in the linear motor, there is a permanent magnetic field in addition to the magnetic field from the current.

This is called the linear motor since the motion of the armature is linear. The Fig. 5.5 shows a DC linear motor which is very similar to the railgun, where there are two rails and a magnetic field perpendicular to the rails.

Fig. 5.5 The DC linear induction motor

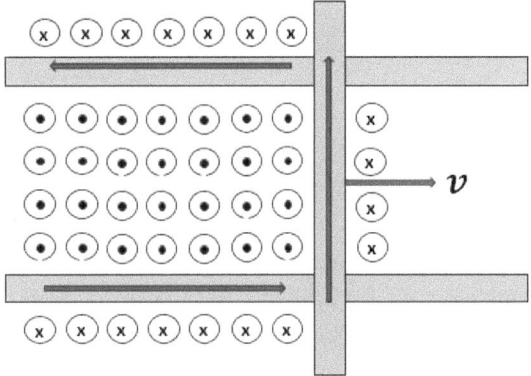

The rails are connected by a moving armature. The armature is made of a conducting metal. Let's consider the length of the armature to be l. There is current flowing in the rails as well as the armature. In the case of the railgun, we studied the force acting on the projectile. In this case, let's study the potential difference in the armature. Let's start with the Lorentz force:

$$F = q(E + v \times B) \tag{5.6}$$

If the armature moves at a constant speed, the total force acting on it should be zero. Hence:

$$E = -v \times B \tag{5.6a}$$

This electric field is induced due to the motion of the armature. The total voltage developed the across the armature is:

$$E = -\nabla\Phi \tag{5.6b}$$

$$\Phi = (v \times B)l \tag{5.6c}$$

If the supply voltage is V_{DC}, and the resistance of the rails is R_{rail}, the current through the circuit is:

$$I = \frac{V_{DC} - \Phi}{R_{rail}} \tag{5.6d}$$

Also as shown earlier, the force on the armature or the loading can be calculated as:

$$F_m = Il \times B \tag{5.6e}$$

Reference

- Website: https://slideplayer.com/slide/12311054/ accessed June 25, 2024.

Application 50: Three-Phase Linear Motor

The three-phase linear motor works like the DC linear motor—but this time there are multiple coils fed with the three-phase currents. It is worth mentioning that there are multiple linear motor designs in the industry and only one design type is explained here to explain the basics. The Fig. 5.6 shows a three-phase linear motor. Here the moving armature consists of three coils. In a typical arrangement, there can be any number of electromagnets in the armature, although only three are shown in the figure. Each electromagnet is fed with a phase from the three-phase current. Based on the phase of the current, each electromagnet will induce either a north or a south pole at the facing end of the permanent magnets. The permanent magnets are arranged in opposite polarities.

The three-phase linear motor works as follows: when there are like poles facing each other they impose a repelling force. At the same time, the adjacent opposite pole of the permanent magnet attracts the electromagnet. This pushing and pulling action creates a total force towards the left as shown in the figure—making the armature move forward.

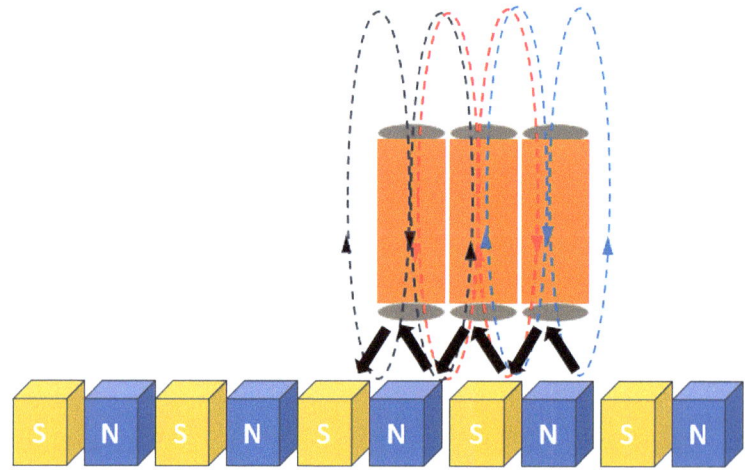

Fig. 5.6 Three phase linear motor

Application 51: Maglev Train

Maglev stands for magnetic levitation. Currently, there are multiple magnetic levitated trains in the world and here we will be discussing about the Japanese SC-Maglev train the fastest train in the world. This train can achieve a speed of 603 km/h. The letters SC in the name stands for superconducting since this train uses superconductor loops along the length of the train to work as electromagnets. Superconductors can conduct heavy currents at zero resistance. Theoretically, once charged these superconductors can conduct current for an infinitely long time. Practically, due to eddy currents there is some heat generated, but still these superconductors can conduct several hundred kilo amperes for a few hours. The superconductors used in the SC-maglev train conduct a current of 700kA. These superconductive loops are arranged along the length of the train such that the adjacent loops carry currents in opposite directions. Let's consider the interactions of these super-conducting electromagnets with the propulsion, levitation, and guidance magnets.

The propulsion is achieved by regular electromagnets, arranged along the sides of the side rails (Fig. 5.7). These loop-electromagnets carry currents in opposite directions such that alternating north and south poles face the electromagnets of the train. These alternating north and south poles of the propulsion coils impose repelling and attracting forces on the superconducting electromagnets along the sides of the train propelling the train forward.

The levitation of the train is achieved by a square loop inter-twined in the shape of figure 8 as shown in Fig. 5.8. These coils are not connected to a power source. As the train passes along the side rails, the super-conducting electromagnets along the sides of the train induce an electromotive force on these coils. Since the coil is inter-twined it acts as two coils. When the train is centered on the coil, the top and bottom parts of the inter-twined coil act as two magnets with equal strength oriented in the opposite direction hence the net electromotive force on the coil is zero and there will be no current

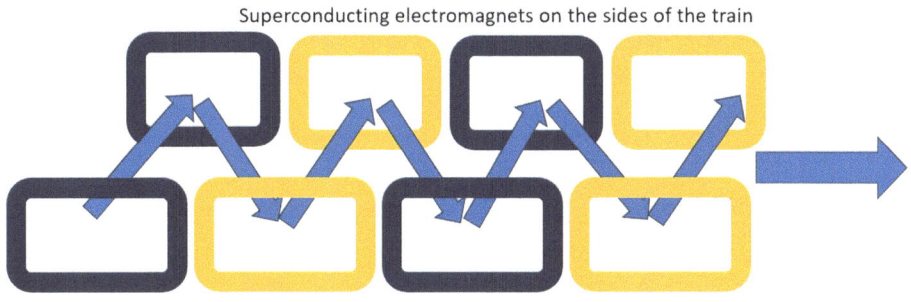

Superconducting electromagnets on the sides of the train

Electromagnets on the side-rails

Fig. 5.7 The propulsion mechanism of magnetically levitated train

Fig. 5.8 The levitation and guidance mechanism of the Maglev train

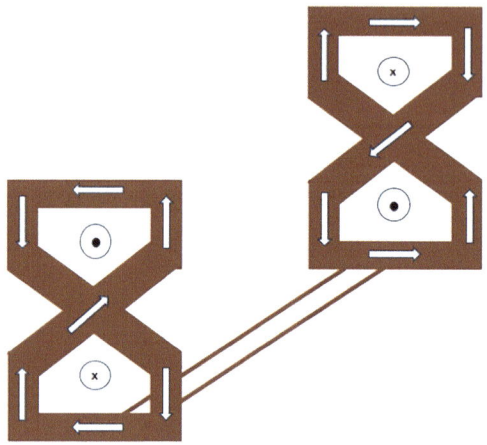

through the loop. When the train is not levitated, the strength of the magnetic field in the bottom half of the inter-twined loop is higher than the top half. Hence there will be a net electromotive force on the coil creating a current. The direction of this current is such that, it induces a magnetic field on the top half of the loop to attract the super-conducting magnet of the train while the bottom half of the inter-twined loop repels it. Once this attraction and repelling force is equal to the weight of the train, it starts levitating.

In addition to being levitated, the train must be centered in the guided railway as well. This is achieved by the interconnected loops of the figure 8-shaped coils. When the train is not centered, the electromotive force acting on the figure 8-shaped coil on one side of the side rail will be different from the other. This difference creates a current through the interconnected loops. Remember the electromotive force is a voltage and the voltage difference creates a current. This current increases the electromotive force on the opposite side of the Fig. 5.8 coil, balancing the EMFs on both sides and making the train centered along the guideway.

It is worth noting that the levitation and the guidance of the maglev require an induced electromotive force. According to Faraday's law, for the EMF to be generated there should be a rate of change of magnetic flux. This changing magnetic flux is produced by the movement of the train. The levitating EMF cannot be generated when the train is at rest. Therefore, the maglev trains have wheels for the initial start-up and the slow-down phases.

References

- Website: https://www.youtube.com/watch?v=XjwF-STGtfE accessed June 25, 2024.
- Website: https://www.globalrailwayreview.com/article/136627/superconducting-mag lev-speeding-toward-sustainability/ accessed June 25, 2024.
- Website: https://link.springer.com/article/10.1007/s40864-019-0104-1 accessed June 25, 2024.

Application 52: Inductive Power Collection (IPC) of Maglev Trains

The magnetically levitated trains consume a lot of power for on-board train cooling systems as well as charging of the side rail electromagnets. To cater to some of these power requirements the Maglev trains use the inductive power collection mechanism. The inductive power collection is the same process as the wireless power transfer. The energized ground coils induce power on the power collection coils or the charging coils on the floor of the train. This power can be used on board the train.

Application 53: AC Synchronous Generator or the Alternator

The alternator works based on Faraday's law. In this case, the induced electromotive force is called Motional EMF since the loop is rotating. An alternator or an AC generator is composed of a conductive loop with one or multiple turns called the rotor, slip rings, and a permanent magnetic field. In an alternator, each end of the loop is permanently connected to one of the rings, and the voltage/current is measured across the rings. The goal of an alternator is to convert rotational or mechanical energy to electrical energy. As the rotor rotates, the magnetic flux through it changes inducing an electromotive force. The direction of the current generated due to the electromotive force is given by Fleming's right-hand rule. If the rotor rotates clockwise, as shown in Fig. 5.9a, the force of the AB edge of the rotor is upwards, hence the current will be into the board. In the second orientation, the magnetic field and the force are parallel to each other hence the induced current is zero. In the third configuration shown in Fig. 5.9c, the force on the AB edge of the rotor is downward, hence the current is out of the board—opposite to the direction in Fig. 5.9a. The overall voltage/current generated by an alternator is an alternative hence its name.

In the alternator, the rotor rotates at a constant angular frequency, and the induced EMF and the current are proportional to the angular frequency of the rotor. Hence the alternator is a synchronous generator.

The force acting on the alternator arm is given by:

$$F = qvBsin\theta \tag{5.7}$$

Or

$$F = IlBsin\theta \tag{5.7a}$$

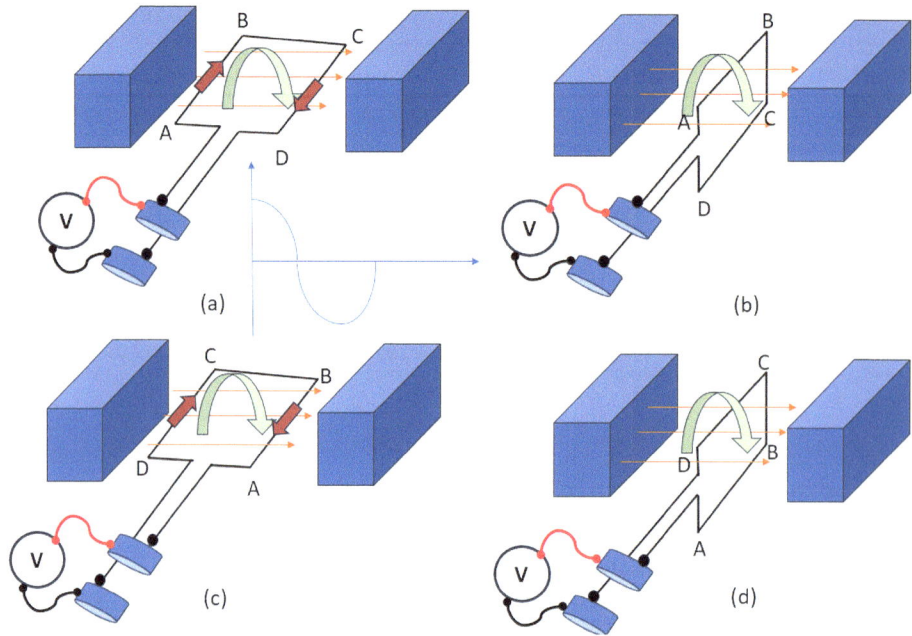

Fig. 5.9 The stages of the AC synchronous generator

Application 54: DC Synchronous Generator

The DC synchronous generator is quite like the alternator, with the only difference being, that the ends of the loop are not connected to the rings. Instead, as the rotor rotates, the two ends switch their connection from one ring to the other as shown in Fig. 5.10. These rings are called commutator rings or slip rings. Hence the direction of voltage or current generated between the two commutator rings always stays in the same direction.

Fig. 5.10 DC synchronous generator

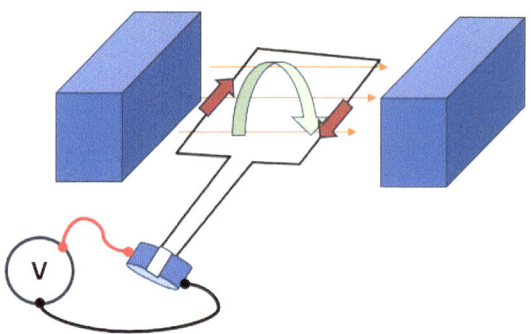

Application 55: The DC Motor

The simplest type of a DC motor consists of two permanent magnets(stator), and a coil powered by direct current(rotor) as shown in Fig. 5.11a. Permanent magnets are used in the smallest DC motors and larger motors use electromagnets powered by DC currents. The configuration of a DC motor is very similar to a DC generator, except in the motor, the coil is connected to power. When a current-carrying conductor is placed in a magnetic field, according to the Lorentz Law there is a force acting on the current-carrying conductor—forcing the conductor to rotate. The two ends of the conductor are connected to commutator rings.

The Force acting on one edge of the loop can be calculated as:

$$F = qvBsin\theta \tag{5.8}$$

$$F = IlBsin\theta \tag{5.8a}$$

In Eq. 5.8a, l is the arm length of the loop and θ is the angle between the magnetic field and the current.

The torque produced by the loop is:

$$T = IldBsin\theta \tag{5.8b}$$

One issue of having only one loop is, that when the loop is at its vertical orientation, the force acting on it is zero. Hence there is a glitch in rotation. That issue can be avoided by having two loops perpendicular to each other as shown in Fig. 5.11b. Most of the DC motors have multiple loops encased in iron slots. Having multiple loops makes the rotation of the motor smooth and the iron slots increase the interaction with the magnetic field. The DC motors are not self-starting.

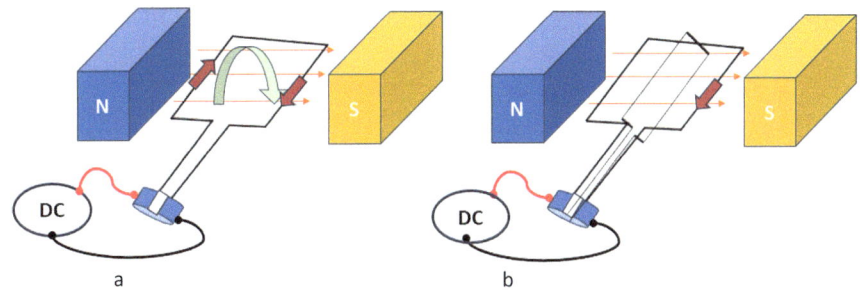

Fig. 5.11 **a** (Left) a DC motor with only one loop. **b** (Right) a DC motor with two perpendicular loops

References

- Website: https://www.youtube.com/watch?v=LAtPHANEfQo&t=121s accessed June 25, 2024.
- Website: https://homepages.laas.fr/lzaccari/seminars/DCmotors.pdf accessed June 25, 2024.

Application 56: Back EMF Calculation

By far we know from Faraday's law, that a time-varying magnetic field through a loop can induce a voltage across its ends. This voltage is called the electromotive force and the current produced by the EMF opposes the change caused by the original magnetic field by creating a secondary magnetic field.

Now, let's consider the case of the DC motor, where there is at least one loop in a magnetic field. The DC motor is not self-starting since it needs an initial current to make the loop spin. But once the loop starts spinning it generates an electromotive force according to Faraday's Law. In other words, the motor starts showing a generator behavior. This induced electromotive force of a motor is called back EMF. The back EMF can be calculated as follows:

$$EMF = \oint_l \boldsymbol{E}.d\boldsymbol{l} = -\frac{\partial \phi_m}{\partial t} \tag{5.9}$$

$$EMF = -\frac{\partial(|\boldsymbol{B}|A\cos\theta(t))}{\partial t} \tag{5.9a}$$

In the above equation, the angle θ is given as a function of time, because in a DC motor, the magnetic field ($\boldsymbol{B}$) and the cross-sectional area of the loop (A) stay the same. The only thing that changes concerning time is the angle of the loop concerning the magnetic field. The differentiation of the above equation with respect to time yields:

$$EMF = |\boldsymbol{B}|A\frac{\partial\theta(t)}{\partial t}\sin(\theta(t)) \tag{5.9b}$$

$\frac{\partial\theta(t)}{\partial t}$ is the angular velocity of the loop $\omega(t)$. Therefore, the above equation becomes:

$$EMF = |\boldsymbol{B}|A\omega(t)\sin(\theta(t)) \tag{5.9c}$$

Is the back EMF desirable?

The answer to this question is a yes. Because without the back EMF, the current through the loop can increase indefinitely as the loop spins through the magnetic field.

Due to the back EMF, the current cannot increase indefinitely. This makes the DC motor a self-regulated machine.

Application 57: The 3-phase AC Induction Motor

The AC induction motor is currently the most widely used industrial motor. As the name implies, this motor is powered by three phases of alternating current. This three-phase alternating current generates a rotating, time-varying magnetic field. As we know already according to Faraday's law of induction, when a conducting loop is placed on a time-varying magnetic field, there is an electromotive force produced in that loop. This electromotive force generates a current through the loop.

Now, let's combine the Faraday's law with the Lorentz law. The Lorentz law says, that when there is a current and a perpendicular magnetic field it produces a magnetic force. This magnetic force can be written as at a given time instant. Remember here, that the magnetic field and the current are time-dependent variables.

$$F = IlBsin\theta \hspace{3cm} (5.10)$$

In the above equation, I is the current through the loop, l is the length of the arm B is the magnitude of the magnetic field and θ is the angle between the current and the magnetic field. This force created due to the induced current makes the rotor spin. Hence the name induction motor. The induction motors are self-starting. Unlike the DC motor, the induction motor does not require the rotor directly connected to the power. Instead of having one or multiple loops, the induction motor's rotor is built like a squirrel cage with short ends. Figure 5.12a shows the stator configuration of an AC motor and the side-view of the squirrel cage rotor Fig. 5.12b.

The rotational speed of the magnetic field is called the synchronous speed N_s, and the rotational speed of the rotor is called N_{rotor}. When these two speeds are equal to each other the loops do not cut the magnetic field hence the force generated on the rotor bars is zero. This makes the rotor slow down. As the rotor slows down, the rotor starts cutting the magnetic field inducing EMF and in turn force. The difference between the synchronous speed and the rotor speed is known as the slip.

$$slip = \frac{N_s - N_{rotor}}{N_s} \times 100\% \hspace{2cm} (5.10a)$$

The most important feature of the induction motor is that the synchronous speed of the rotating magnetic field is proportional to the frequency of the input power and the rotor always tries to catch up to the synchronous speed. Hence, by changing the frequency of the input power, the rotor speed of the AC motor can be controlled.

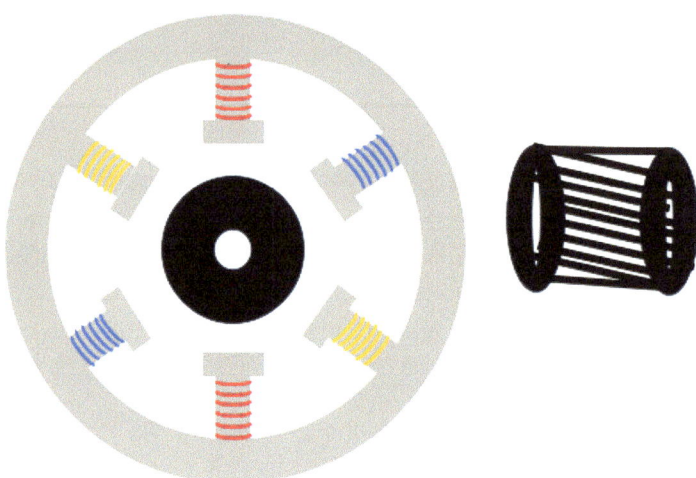

Fig. 5.12 a (Left) the stator windings of an AC motor and the side view of the squirrel cage rotor (right)

Reference

- Website: https://www.youtube.com/watch?v=AQqyGNOP_3o&t=312s accessed June 25, 2024.

Application 58: Synchronous Reluctance Motor

The stator configuration of a synchronous reluctance motor is similar to the AC induction motor. The main difference lies in the rotor. The AC induction motor uses conducting loops or a squirrel cage structure as the rotor. However, the reluctance motor uses iron-like material with high permeability. The permeability is the ability of a material to generate internal magnetic flux lines in response to an external magnetic field. These materials can align themselves along the external magnetic fields. The synchronous reluctance motor uses this characteristic of ferromagnetic materials and uses those as rotors. Figure 5.13 shows the cross section of a synchronous reluctance motor.

As the magnetic field generated by the three-phase current rotates, the ferromagnetic bar tries to stay aligned with the rotating magnetic field. Due to the inertia of the bar and the frequency of the input current, proper rotation of the bar was not achieved. This issue was eliminated by having a +-shaped ferromagnetic rotor. Interested reader is suggested to refer to the references to gain a deeper understanding of this motor.

Fig. 5.13 Cross section of a
synchronous reluctance motor

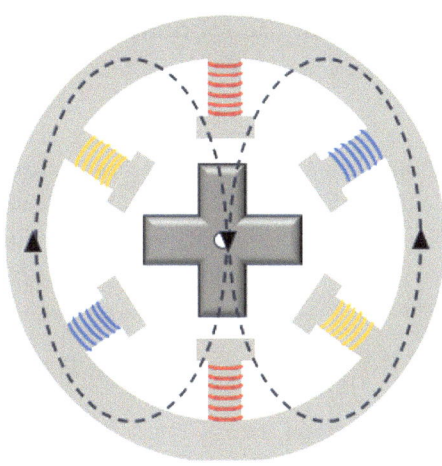

References

- Website: https://www.youtube.com/watch?v=vvw6k4ppUZU accessed June 25, 2024.
- Website: https://pnbalamurugan.yolasite.com/resources/EE6703%20SEM%20UNIT%201%20-%20SYNCHRONOUS%20RELUCTANCE%20MOTORS.pdf accessed June 25, 2024.

Application 59: Universal Motor

A universal motor is designed to work either with DC or single-phase AC currents. The operational principle of this motor is the same as the DC motor above. Although a DC motor can use either permanent magnets or electromagnets to produce the magnetic field, universal motors can only use electromagnets. Under DC currents these motors will work the same way as a DC motor. When connected to an AC, it is expected that the direction of the current changes with the phase of the input current. The important thing is when the direction of the current changes, the direction of the magnetic field produced by the electromagnets changes as well, keeping the force in the same direction. Therefore, the direction of rotation stays the same. These motors require additional lamination and commutator ring brushes to reduce eddy current losses and sparking.

Reference

- Website: https://www.youtube.com/watch?v=0PDRJKz-mqE accessed June 25, 2024.

Application 60: Modelling Particle Motion in a Plasma

Plasma is considered the fourth state of matter where many of the particles are at an ionized state. The interesting thing about plasma is that 99% of the universe is in a plasma state including the Earth's ionosphere and the magnetosphere.

Given that plasma contains charged particles, the motion of these particles in a plasma is modeled by the Lorentz force. Modelling the plasma particle is important to understand the physics of the universe, in space satellite communications as well as in building plasma devices.

$$F = q(E + v \times B) \tag{5.11}$$

The nominal state of a plasma is quasi-equilibrium. But given the presence of charges and their interactions with the waves, there will be electric fields and magnetic fields. In the case of Earth's near space, the dominant magnetic field is the geomagnetic field. Let's consider the components of the Lorentz force step by step.

I. $E \neq 0, B = 0$

If there is only an electric field present the Lorentz force equation simplifies to:

$$F = qE \tag{5.11a}$$

Given the above force, the positive ions will show a linear motion in the direction of the electric field, and the electrons and negative ions will show a linear motion in the opposite direction of the electric field.

II. $E = 0, B \neq 0$

This scenario can be analyzed in two parts.

a. For simplicity let's assume that the magnetic field is oriented in the a_z, direction and the particle velocities are in the a_y direction. The resulting Lorentz force will be:

$$F = q(v_y a_y \times B_z a_z) = q v_y B_z a_x \tag{5.11b}$$

This curl operation will create a circular motion of the particles. The frequency of this circular motion is called the cyclotron frequency or the gyrofrequency and its magnitude is given by:

$$|\omega_c| = \frac{|q|B}{m} \tag{5.11c}$$

In Eq. 5.11c, q is the charge, B is the magnitude of the magnetic field and m is the mass. Given that electron and ion masses differ electrons have a higher cyclotron frequency compared to ions.

b. Now, let's consider the case where the particle velocity has a parallel and a perpendicular component to the magnetic field.

$$v = v_y a_y + v_z a_z \qquad (5.11d)$$

In this case, the Lorentz force $qv_y B_z a_x$ will force the circular motion of the particles. At the same time due to the parallel component $v_z a_z$, the particles move linearly along the magnetic field lines. Hence the combined motion will be a helix. This is the typical motion of particles trapped in Earth's radiation belts.

III. $E \neq 0, B \neq 0$

When both the electric and magnetic fields are present, the particle motion becomes complicated since it starts showing a hybrid of linear and helical motion in the resultant direction.

In addition to the above basic scenarios, there are other types of particle motion such as the curvature drift of the magnetic field and the gradient drift.

Reference

- Bittencourt, J. (2003) Fundamentals of Plasma Physics. 3rd Edition, Springer, Berlin.

Application 61: Plasma Propulsion Systems—RF-Heated Electrothermal Thruster

Plasma thrusters are widely used in spacecraft due to their fuel efficiency. The simplest type of a plasma thruster is the electrothermal thruster. A simple electrothermal thruster is comprised of a radio frequency antenna, a sheath, and an exhaust valve as shown in Fig. 5.14.

When there is an alternating current running in the antenna, according to Faraday's law, it induces an electromotive force or a voltage. This voltage ionizes the gaseous propellant at low pressure by stripping away its electrons, hence creating a plasma. This process is known as ionization by induction and it's an electrodeless mechanism. This RF energy also heats the plasma particles, which in turn increases the kinetic energy of the particles providing the thrust to the spacecraft. The exit velocity of the propellant is approximated by:

$$v_{exit} \leq \sqrt{2c_p T_c} \qquad (5.12)$$

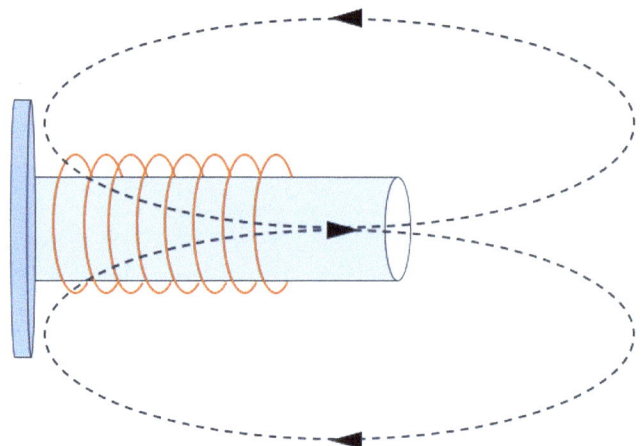

Fig. 5.14 RF heated electrothermal thruster

In Eq. 5.12, c_p is the propellent at constant pressure per unit mass and T_c is the maximum temperature that can be withstood by the device.

References

- Website: https://fti.neep.wisc.edu/fti.neep.wisc.edu/_jfs/neep533.lect31.99/plasmaProp. html accessed June 25, 2024.
- Website: http://currentpropulsionsystems.weebly.com/electromagnetic-propulsion-sys tems.html accessed June 25, 2024.
- Bathgate, Stephen & Bilek, Marcela & MCKENZIE, D. (2017). Electrodeless plasma thrusters for spacecraft: A review. Plasma Science and Technology. 19. 083,001. https:// doi.org/10.1088/2058-6272/aa71fe.

Application 62: Electrostatic Thrusters—Ion Thruster

Electrostatic thrusters (shown in Fig. 5.15) use electric fields to generate the thrust. In an ion thruster an electron gun or a hollow cathode bombard the electron onto the propellent (commonly Xenon due to its high atomic number). The bombardment of electrons creates positive Xenon ions and releases electrons. The magnets on the walls of the reactor lengthen the interaction time between the electrons and the propellent by preventing them from reaching the discharge chamber walls.

These positively charged *Xe* ions diffuse towards an even more positively charged screen grid due to the pressure build-up within the chamber. The purpose of the screen

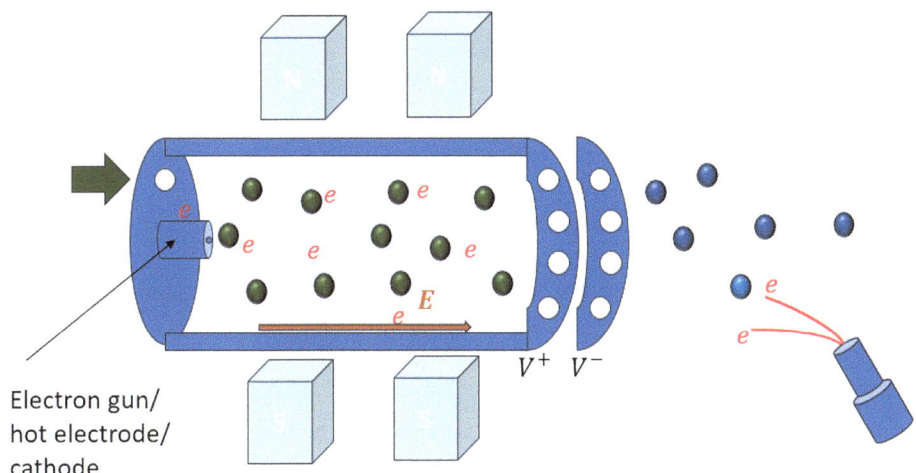

Fig. 5.15 A schematic of an ion-thruster

grid is to prevent heavy propellant ions from bombarding the accelerator grid (second grid). The holes of the screen grid and the accelerator grid are very precisely aligned. As the propellant ions pass through the screen grid they get accelerated towards the negatively charged accelerator grid at a velocity of up to 90000mph providing the thrust. The electrons may pass through the screen grid, but they are repelled by the negatively charged accelerator grid pushing them back into the chamber.

Once the ions exit the thruster as an ion beam, a neutralizer is used to emit electrons and neutralize the positive propellent ions. Without this neutralization, the body of the spacecraft gets negatively charged and the positively charged positive ions attract back to the spacecraft reducing the thrust.

The thrust or the force created by the electrostatic ion thruster is given by:

$$F_{thrust} = qE \tag{5.13}$$

In Eq. 5.13 q is the total charge of Xe ions and E, is the electric field intensity inside the discharge chamber.

References

- Website: https://www.nasa.gov/wp-content/uploads/2015/08/ionpropfact_sheet_ps-01628.pdf accessed June 25, 2024.
- Website: https://descanso.jpl.nasa.gov/SciTechBook/series1/Goebel_05_Chap5_Grids.pdf accessed June 25, 2024.

- Website: https://www.youtube.com/watch?app=desktop&v=grU8g9jnS4w   accessed June 25, 2024.
- Website: https://www.youtube.com/watch?v=bHunhXk9i2s&t=393s accessed June 25, 2024.

Application 63: Hall-Effect Thruster

Hall effect thruster (shown in Fig. 5.16) is considered an electrodynamic thruster. In this case, there is an electric field applied in the direction shown and a magnetic field perpendicular to the electric field. The cathode releases electrons that are trying to reach the anode. From the anode end, there is an influx of Xenon atoms. If there was only an electric field present the behavior of this thruster would be the same as the ion thruster. But the difference comes from the Hall-current. Due to the applied magnetic field, these electrons are deviated from their linear motion and forced to gyrate inside the discharge chamber—trapping them inside. The current generated due to this gyration is called the Hall current.

These trapped electrons interact with the noble Xe atoms, it produces positive Xe ions and more electrons. These Xe ions are then pushed out of the thruster by the Lorenz force.

$$F_{thrust} = qE \qquad (5.14)$$

Like the ion thruster, a neutralizer can be used to neutralize the exiting Xe ions to avoid spacecraft charging.

Why do only the electrons gyrate and not the Xe ions? Both the electrons and Xe ions can gyrate. But the gyro radius of electrons is much smaller than the gyro radius of Xe ions. The radius of the discharge chamber is designed such that it is greater than the gyro radius of electrons but much smaller than the gyro radius of Xe ions. Hence the motion of Xe ions is almost linear.

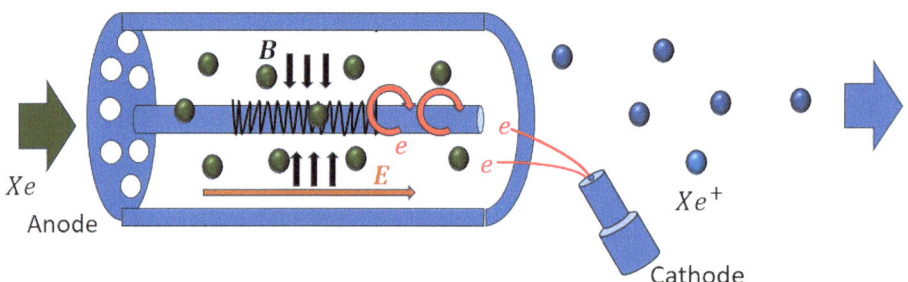

Fig. 5.16 A schematic of a Hall-effect thruster

References

- Website: https://technology.nasa.gov/patent/LEW-TOPS-34 accessed June 2024.
- Website: https://www.youtube.com/watch?v=mAfjmGMp43w accessed June 2024.

Application 64: Magneto Plasma Dynamic (MPD) Thruster

Magneto plasma dynamic thruster is also an electro-dynamic thruster. In this case, there is a cylindrical cathode and an anode surrounding it. The electrons are emitted by the cathode bombards on the Xe atoms releasing more electrons and creating positive Xe ions. The Fig. 5.17 shows the direction of the conventional current and the electron current would be in the opposite direction.

As there is a magnetic field generated around every current-carrying conductor, here also, a self-induced magnetic field will also be generated according to the right-hand rule. The cross-product between the current and the magnetic field generated a magnetic Lorentz force. This Lorentz force acts on the electrons in the right-hand direction since the electrons are easier to push due to their low mass. As the electron exits the thruster, the attracted Xe ions are pushed with them creating the thrust.

The thrust generated this way is not that high. Hence, to increase the thrust an electro-magnetic field is applied in the middle of the thruster. This is known as the applied field MPD. But these electromagnets require a high amount of power that can be generated only by nuclear reactors.

$$F_{thrust} = qv \times B \qquad (5.15)$$

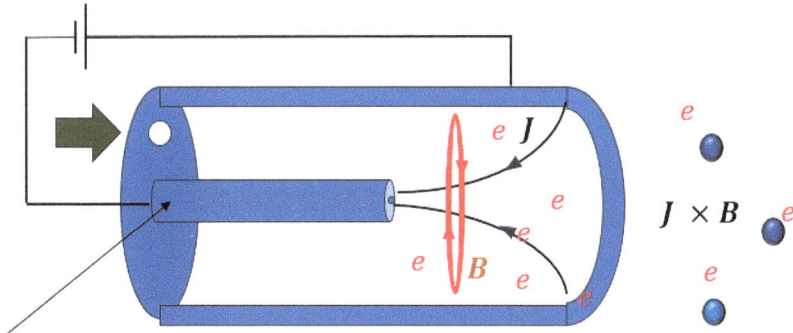

Electron gun/hot electrode/ hollow cathode

Fig. 5.17 The schematic of an electrodynamic thruster

References

- Website: https://www.youtube.com/watch?v=M3iYhtsZCiA, accessed June 25, 2024.
- Website: https://www.daviddarling.info/encyclopedia/M/MPDthruster.html#google_vig nette, accessed June 25, 2024.

Application 65: Helicon Thruster

Helicon thruster is an electrodeless plasma thruster mechanism. The thruster gets its name due to the helicon antenna used to energize the plasma (Fig. 5.18). The thruster with only a Helicon antenna is considered an electrothermal thruster. To improve the efficiency in particle acceleration a diverging magnetic field is used. The purpose of the two coils is to create a diverging magnetic field. This diverging magnetic field, also known as a magnetic nozzle, creates a difference in electron concentration creating a potential difference or a voltage. This diverging magnetic field and the resulting electric field gradient are used to create a Lorentz force on the electrons such that they are pushed out of the thruster carrying the heavy propellant atoms with them. This creates the thrust. The generated thrust can be calculated by:

$$F_{thrust} = qv \times B \tag{5.16}$$

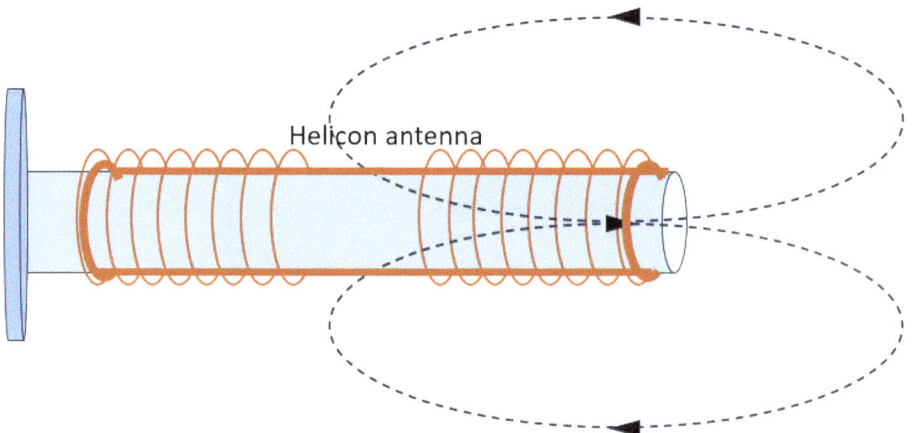

Fig. 5.18 A schematic of a helicon thruster

References

- Bathgate, Stephen & Bilek, Marcela & MCKENZIE, D. (2017). Electrodeless plasma thrusters for spacecraft: A review. Plasma Science and Technology. 19. 083001. https://doi.org/10.1088/2058-6272/aa71fe.
- Melazzi, Davide, et al. "Antenna design and optimization for helicon plasma thruster with coupled surface and volume integral equations." *2013 International Conference on Electromagnetics in Advanced Applications (ICEAA)* (2013): 712–715.
- Website: https://www.youtube.com/watch?v=gHdmxd459ZY, accessed June 25, 2024.

Practice problems.

1. In Earth's magnetosphere, high energy electrons are spinning around the geomagnetic field, in regions known as radiation belts.

    Without losing generality, let's assume that the geomagnetic field at the point of observation is $50\mu T a_z$, and the velocity of a particle is $2.5 \times 10^8 a_y + 2.5 \times 10^8 a_z$ ms^{-1}.

    a. Calculate the motional electric field intensity (Em) and determine its direction.

    Hint: $E_m = -v \times B$.

    b. If a satellite contains a rectangular loop receiving antenna with a length of 1m and width of 75 cm, how much of a motional electro-motive force is generated within the loop antenna? Here you need to calculate only the magnitude of it

    Hint: motional emf $= \oint E_m.dl$

    c. If there are 106 particles in the region of interest and all particles are electrons, using Lorentz's force formula, calculate the total force.

    Hint: $F = Qv \times B$

    d. Using the velocity given, calculate the current within 1 m in the magnetosphere parallel and perpendicular to the geomagnetic field. Hint: current = charge × velocity.

    e. Using the velocity given, calculate the current within 1m in the magnetosphere parallel and perpendicular to the geomagnetic field. Hint: current = charge x velocity.

2. Inside a dynamo, a wire loop or a coil rotates within a magnetic field. In an actual dynamo the wire loop consists of multiple turns, but here let's consider a single loop. In this case, consider the rectangular wire loop (15 cm × 10 cm).

    a. The electromotive force produced by the wire loop is 0.075 V. How much is the electric field intensity within the loop?

    b. The strength of the magnetic flux density is 0.01 T (Tesla). The loop spins at a rate of 50 rotations per second. Calculate the rate of change of magnetic flux within the loop.

   c. If we want to generate an emf of 5 V using the same loop and the same magnet, how much should be the rotation rate of the wire loop?

   d. How much is the new electric field intensity within the loop when the emf is 5V?

3. Magnetic flow meter.

   a. The separation between the electrodes is 10 cm. if the magnitude of the electric field intensity between the electrodes is 30 V/m, calculate the induced electromotive force (EMF) between the electrodes.

   b. The magnitude of the magnetic flux density inside the device is 0.2 T (B), and the conductor length is 10 cm (d: the separation between the electrodes). Using the EMF calculated in part a, calculate the velocity (flow rate). Hint: pay attention to the units. $V = Tm2/s = Wb/s$.

   c. If the liquid flow rate is increased to 200 m/s how much would be the new induced potential across the electrodes?

   d. The minimum measurable voltage between the electrodes is 0.1 V. The separation between the electrodes is 10 cm and the minimum magnetic flux density is 0.1 T. What is the minimum detectable flow rate?

Electromagnetic Waves and Electromagnetic Power

6

This chapter discusses the applications of Maxwell's equations in electromagnetic waves and electromagnetic power. Electromagnetic waves are one of the reasons why Maxwell's equations are considered revolutionary.

Application 66: Plane Wave Solution

The plane wave solution demonstrates the wave propagation in the far field. This equation is important for long-range wireless communications, including cell phone, satellite, and radar communications.

Let's start with the point from Maxwell's equations.

$$\nabla . \boldsymbol{D} = \rho_v \tag{6.1a}$$

$$\nabla . \boldsymbol{B} = 0 \tag{6.1b}$$

$$\nabla \times \boldsymbol{E} = -\frac{\partial \boldsymbol{B}}{\partial t} \tag{6.1c}$$

$$\nabla \times \boldsymbol{H} = \sigma \boldsymbol{E} + \frac{\partial \boldsymbol{D}}{\partial t} \tag{6.1d}$$

The equations above show the strongly coupled electric and magnetic fields. To solve either for electric or magnetic fields we need to decouple the above fields. Let's take Maxwell's 3rd equation and take the curl of both sides once more.

© The Author(s), under exclusive license to Springer Nature Switzerland AG 2025 109
A. Maxworth, *One Hundred Applications of Maxwell's Equations*, Synthesis Lectures on Electromagnetics, https://doi.org/10.1007/978-3-031-73784-8_6

$$\nabla \times (\nabla \times \boldsymbol{E}) = \nabla \times \left(-\frac{\partial \boldsymbol{B}}{\partial t}\right) \tag{6.1e}$$

Given that the electric and magnetic fields are continuously differentiable, we can switch the order of the spatial and time derivatives.

$$\nabla \times (\nabla \times \boldsymbol{E}) = -\frac{\partial (\nabla \times \boldsymbol{B})}{\partial t} \tag{6.1f}$$

$$\nabla \times (\nabla \times \boldsymbol{E}) = -\mu \frac{\partial (\nabla \times \boldsymbol{H})}{\partial t} \tag{6.1g}$$

$$\nabla \times (\nabla \times \boldsymbol{E}) = -\mu \frac{\partial}{\partial t}\left(\sigma \boldsymbol{E} + \frac{\partial \boldsymbol{D}}{\partial t}\right) \tag{6.1h}$$

$$\nabla \times (\nabla \times \boldsymbol{E}) = -\mu\sigma \frac{\partial \boldsymbol{E}}{\partial t} - \mu\varepsilon \frac{\partial}{\partial t}\left(\frac{\partial \boldsymbol{E}}{\partial t}\right) \tag{6.1i}$$

Now, let's use a vector identity for the left-hand side of the above equation.

$$\nabla \times (\nabla \times \boldsymbol{E}) = \nabla(\nabla.\boldsymbol{E}) - \nabla^2 \boldsymbol{E} \tag{6.1j}$$

Let's find $\nabla.\boldsymbol{E}$, using the Eq. 6.1a $\nabla.\boldsymbol{D} = \rho_v$.
If the region is source-free:

$$\nabla.\boldsymbol{D} = 0 \tag{6.1k}$$

$$\nabla.(\varepsilon \boldsymbol{E}) = 0 \tag{6.1l}$$

$$\nabla\varepsilon.\boldsymbol{E} + \varepsilon\nabla.\boldsymbol{E} = 0 \tag{6.1m}$$

$$\nabla.\boldsymbol{E} = -\boldsymbol{E}\frac{\nabla\varepsilon}{\varepsilon} \tag{6.1n}$$

$\nabla.\boldsymbol{E}$ is nonzero, if and only if there is a permittivity gradient. In long-range electromagnetic propagation, we consider a constant permittivity. Hence, $\nabla.\boldsymbol{E} = 0$. In optical communications, the graded index fiber optic cable is designed with a gradient in permittivity.

Hence, we can get the homogeneous form of the equation.

$$\nabla^2 \boldsymbol{E} - \mu\sigma \frac{\partial \boldsymbol{E}}{\partial t} - \mu\varepsilon \frac{\partial}{\partial t}\left(\frac{\partial \boldsymbol{E}}{\partial t}\right) = 0 \tag{6.1o}$$

$$\nabla^2 \boldsymbol{H} - \mu\sigma \frac{\partial \boldsymbol{H}}{\partial t} - \mu\varepsilon \frac{\partial}{\partial t}\left(\frac{\partial \boldsymbol{H}}{\partial t}\right) = 0 \tag{6.1p}$$

At this point, we separate the spatial and time derivatives assuming that the fields are time-harmonic. For example, we believe that the wave is propagating in the z direction. The E field is oriented in the a_x direction.

$$E(z,t) = E_s(z)e^{j\omega t}a_x \tag{6.1q}$$

$$\nabla^2 = \frac{\partial^2}{\partial x^2} + \frac{\partial^2}{\partial y^2} + \frac{\partial^2}{\partial z^2} \tag{6.1r}$$

If the wave is a plane wave, the wavefront is a plane. And there is no variation w.r.t the x and y directions. Hence

$$\frac{\partial^2 E}{\partial x^2} = \frac{\partial^2 E}{\partial y^2} = 0 \tag{6.1s}$$

$$\frac{\partial E}{\partial t} = j\omega E_s(z)e^{j\omega t}a_x \tag{6.1t}$$

$$\frac{\partial}{\partial t}\left(\frac{\partial E}{\partial t}\right) = (j\omega)^2 E_s(z)e^{j\omega t}a_x \tag{6.1u}$$

Now, let's substitute all these in the homogeneous electric field equation.

$$e^{j\omega t}\frac{\partial^2 E_s(z)}{\partial z^2}a_x - \mu\sigma j\omega E_s(z)e^{j\omega t}a_x - \mu\varepsilon(j\omega)^2 E_s(z)e^{j\omega t}a_x = 0 \tag{6.1v}$$

The time-harmonic function and the unit vector are redundant hence we get the scaler wave equation as:

$$\frac{\partial^2 E_s(z)}{\partial z^2} - \mu\sigma j\omega E_s(z) - \mu\varepsilon(j\omega)^2 E_s(z) = 0 \tag{6.1w}$$

$$\frac{\partial^2 E_s(z)}{\partial z^2} - j\mu\omega(\sigma + j\omega\varepsilon)E_s(z) = 0 \tag{6.1x}$$

$$\gamma = \sqrt{j\mu\omega(\sigma + j\omega\varepsilon)} \tag{6.1y}$$

$$\frac{\partial^2 E_s(z)}{\partial z^2} - \gamma^2 E_s(z) = 0 \tag{6.1z}$$

This differential equation produces the solution:

$$E_s(z) = Ae^{\pm\gamma z} \tag{6.1aa}$$

By applying the boundary conditions at $z = 0$, $A = E_o^+$, where E_o^+ is the magnitude of the electric field intensity at $z = 0$.

The real-time electric field intensity can be expressed as:

$$E_s(z) = E_o^+ e^{-\gamma z} + E_o^- e^{+\gamma z} \tag{6.1ab}$$

In the above equation, the first term on the right-hand side is called the forward traveling wave, and the second term is called the reverse traveling wave.

The electric field intensity $\mathbf{E}$ in its full form is:

$$\mathbf{E}(z,t) = E_o^+ e^{-\gamma z} e^{j\omega t} \mathbf{a_x} + E_o^- e^{+\gamma z} e^{j\omega t} \mathbf{a_x} \tag{6.1ac}$$

The form of the electric field shown in Eq. 6.1ac is known as the phasor form of the electric field. The phasor form indicates the magnitude and the phase of the electric field. As shown in Eq. 6.1ac, the electric field of the propagating plane wave is a function of the space (z) and time. Also, it is assumed that the field is time-harmonic which is represented by the $e^{j\omega t}$.

Let's assume that the magnetic field intensity is also time-harmonic. Hence by applying Faraday's Law,

$$\nabla \times \mathbf{E} = -\mu \frac{\partial \mathbf{H}}{\partial t} \tag{6.1ad}$$

$$\nabla \times \mathbf{E} = -j\omega\mu e^{j\omega t} \mathbf{H}_s(z) \tag{6.1ae}$$

$$\mathbf{H}_s(z) = \frac{\gamma}{j\omega\mu} E_o^+ e^{-\gamma z} \mathbf{a_y} - \frac{\gamma}{j\omega\mu} E_o^- e^{+\gamma z} \mathbf{a_y} \tag{6.1af}$$

$$\mathbf{H}(z,t) = \frac{\gamma}{j\omega\mu} E_o^+ e^{-\gamma z} e^{j\omega t} \mathbf{a_y} - \frac{\gamma}{j\omega\mu} E_o^- e^{+\gamma z} e^{j\omega t} \mathbf{a_y} \tag{6.1ag}$$

$$\mathbf{H}(z,t) = H_o^+ e^{-\gamma z} e^{j\omega t} \mathbf{a_y} - H_o^- e^{+\gamma z} e^{j\omega t} \mathbf{a_y} \tag{6.1ah}$$

The magnitude ratio between the electric and magnetic fields is known as the wave impedance.

$$\frac{E_o^+}{H_o^+} = -\frac{E_o^-}{H_o^-} = \eta = \sqrt{\frac{j\omega\mu}{\sigma + j\omega\varepsilon}} \tag{6.1i}$$

The term γ is called the propagation constant, which can be broken down into its real and imaginary components.

$$\gamma = \alpha + j\beta \tag{6.1aj}$$

Hence the Realtime electric and magnetic fields become.

$$Re\{\mathbf{E}(z,t)\} = E_o^+ e^{-\alpha z}\cos(\omega t - \beta z)\mathbf{a_x} + E_o^- e^{+\alpha z}\cos(\omega t + \beta z)\mathbf{a_x} \tag{6.1ak}$$

$$Re\{H(z,t)\} = \frac{E_o^+}{\eta}e^{-\alpha z}\cos(\omega t - \beta z)\mathbf{a_y} - \frac{E_o^-}{\eta}e^{+\alpha z}\cos(\omega t + \beta z)\mathbf{a_y} \qquad (6.1\text{al})$$

$$\alpha = \omega\sqrt{\frac{\mu\varepsilon}{2}\left(\sqrt{1+\left(\frac{\sigma}{\omega\varepsilon}\right)^2}-1\right)} \qquad (6.1\text{am})$$

$$\beta = \omega\sqrt{\frac{\mu\varepsilon}{2}\left(\sqrt{1+\left(\frac{\sigma}{\omega\varepsilon}\right)^2}+1\right)} \qquad (6.1\text{an})$$

For free space:

$$\sigma = 0 \qquad (6.1\text{ao})$$

$$\alpha = 0 \qquad (6.1\text{ap})$$

$$\beta = \omega\sqrt{\mu_o\varepsilon_o} = \frac{\omega}{c} = \frac{2\pi}{\lambda} \qquad (6.1\text{aq})$$

$$\eta = \sqrt{\frac{\mu_o}{\varepsilon_o}} = 377\Omega \qquad (6.1\text{ar})$$

For low-loss dielectrics:

$$\frac{\sigma}{\omega\varepsilon} \ll 1 \qquad (6.1\text{as})$$

From the Taylor series expansion for small a value:

$$(1+a)^n = 1 + na \qquad (6.1\text{at})$$

Applying this relationship to

$$\sqrt{1+\left(\frac{\sigma}{\omega\varepsilon}\right)^2} = 1 + \frac{1}{2}\left(\frac{\sigma}{\omega\varepsilon}\right)^2 \qquad (6.1\text{au})$$

Applying this in the equation for the attenuation constant gives:

$$\alpha = \frac{\sigma}{2}\sqrt{\frac{\mu}{\varepsilon}} \qquad (6.1\text{av})$$

$$\beta = \omega\sqrt{\mu\varepsilon} \qquad (6.1\text{aw})$$

$$\eta = \sqrt{\frac{j\omega\mu}{\sigma + j\omega\varepsilon}} \qquad (6.1\text{ax})$$

For electric conductors:

$$\frac{\sigma}{\omega\varepsilon} \gg 1 \tag{6.1ay}$$

$$\alpha = \beta = \sqrt{\frac{\omega\mu\sigma}{2}} \tag{6.1az}$$

$$\eta = \sqrt{\frac{j\omega\mu}{\sigma}} \tag{6.1ba}$$

The loss tangent of a medium is defined as:

$$\tan\delta = \frac{\sigma}{\omega\varepsilon} \tag{6.1bb}$$

The loss tangent is a measure of signal loss.

Application 67: Radio Reflectometry—Transmission and Reflection Coefficients from Normal Incidence on a Dielectric Boundary

Assume an electromagnetic wave enters from one medium to another. At the boundary of the two media, there is an impedance mismatch due to the differences in permittivity and permeability. Because of that portion of the wave transmits into the second medium while a portion reflects to the first medium. The ratio between the transmitted and incident electric field amplitudes is called the transmission coefficient. The reflected electric field amplitude to incidence electric field amplitude is called the reflection coefficient. The transmission and reflection coefficients are important parameters in determining the transmission power and minimizing the return power losses. In this section, we will first derive the transmission and reflection coefficients for the normal incidence of an electromagnetic wave. The directions of the electric and magnetic fields are shown in Fig. 6.1.

Let's consider an electromagnetic wave propagating in the a_z direction. And the boundary is located at $z = 0$. The phasors below represent the incident, transmitted, and reflected electric and magnetic fields.

$$\boldsymbol{E_i}(z, t) = E_o^i e^{j\omega t - (\alpha_1 + j\beta_1)z} \boldsymbol{a_x} \tag{6.2a}$$

$$\boldsymbol{H_i}(z, t) = \frac{E_o^i}{\eta_1} e^{j\omega t - (\alpha_1 + j\beta_1)z} \boldsymbol{a_y} \tag{6.2b}$$

$$\boldsymbol{E_t}(z, t) = E_o^t e^{j\omega t - (\alpha_2 + j\beta_2)z} \boldsymbol{a_x} \tag{6.2c}$$

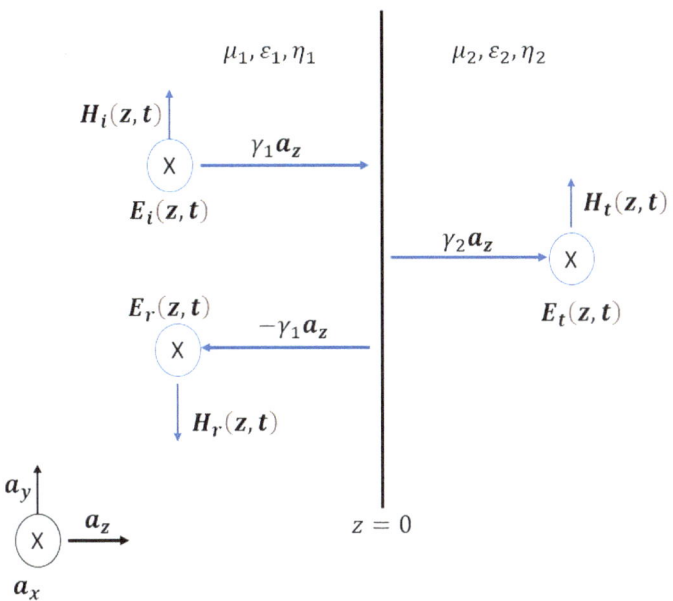

Fig. 6.1 Normal incidence of waves from medium 1 to medium 2 at a boundary located at $z = 0$.

$$H_t(z, t) = \frac{E_o^t}{\eta_2} e^{j\omega t - (\alpha_2 + j\beta_2)z} a_y \qquad (6.2d)$$

$$E_r(z, t) = E_o^r e^{j\omega t + (\alpha_1 + j\beta_1)z} a_x \qquad (6.2e)$$

$$H_r(z, t) = -\frac{E_o^r}{\eta_1} e^{j\omega t + (\alpha_1 + j\beta_1)z} a_x \qquad (6.2f)$$

The real parts of the above phasors are:

$$Re[E_i(z, t)] = E_o^i e^{-\alpha_1 z} \cos(\omega t - \beta_1 z) a_x \qquad (6.2g)$$

$$Re[H_i(z, t)] = \frac{E_o^i}{\eta_1} e^{-\alpha_1 z} \cos(\omega t - \beta_1 z) a_y \qquad (6.2h)$$

$$Re[E_t(z, t)] = E_o^t e^{-\alpha_2 z} \cos(\omega t - \beta_2 z) a_x \qquad (6.2i)$$

$$Re[H_t(z, t)] = \frac{E_o^t}{\eta_2} e^{-\alpha_2 z} \cos(\omega t - \beta_2 z) a_y \qquad (6.2j)$$

$$Re[E_r(z, t)] = E_o^r e^{+\alpha_1 z} \cos(\omega t + \beta_1 z) a_x \qquad (6.2k)$$

$$Re[H_r(z,t)] = -\frac{E_o^r}{\eta_1}e^{+\alpha_1 z}\cos(\omega t + \beta_1 z)a_y \tag{6.2l}$$

Boundary conditions for the electric and magnetic fields at the boundary at $z = 0$.

a.aa ..The tangential electric fields on both sides of the medium should be equal.

b.bb ..The difference between the magnetic field intensities contributes to the surface current density. If the surface current on the boundary is zero, the net magnetic field on both sides of the boundary should be equal.

$$E_o^i + E_o^r = E_o^t \tag{6.2m}$$

$$\frac{E_o^i}{\eta_1} - \frac{E_o^r}{\eta_1} = \frac{E_o^t}{\eta_2} \tag{6.2n}$$

Using the above two equations; the transmission and reflection coefficients of the

$$\tau = \frac{E_o^t}{E_o^i} = \frac{2\eta_2}{\eta_1 + \eta_2} \tag{6.2o}$$

$$\Gamma = \frac{E_o^r}{E_o^i} = \frac{\eta_2 - \eta_1}{\eta_1 + \eta_2} \tag{6.2p}$$

Application 68: Monostatic Radar—Aircraft Surveillance Radar

In radar technology, these reflected fields are called scattered fields. Since the aircraft and the surveillance radar are located far apart, the incidence of the wave can be considered normal to the surface of the aircraft's body. Let's start with the phasor form of incidence and reflected waves. When a field is reflecting from a conductor, $\eta_2 = 0$. And the reflection coefficient $\Gamma = -1$.

$$E_i(z,t) = E_o^i e^{j(\omega t - \beta_1 z)}a_x \tag{6.3a}$$

$$H_i(z,t) = \frac{E_o^i}{\eta_o}e^{j(\omega t - \beta_1 z)}a_y \tag{6.3b}$$

$$E_r(z,t) = E_o^r e^{j(\omega t + \beta_1 z)}a_x \tag{6.3c}$$

$$H_r(z,t) = -\frac{E_o^r}{\eta_o}e^{j(\omega t + \beta_1 z)}a_x \tag{6.3d}$$

The real parts of the above phasers are:

$$Re[E_i(z,t)] = E_o^i \cos(\omega t - \beta_1 z) a_x \tag{6.3e}$$

$$Re[H_i(z,t)] = \frac{E_o^i}{\eta_o} \cos(\omega t - \beta_1 z) a_y \tag{6.3f}$$

$$Re[E_r(z,t)] = E_o^r \cos(\omega t + \beta_1 z) a_x \tag{6.3g}$$

$$Re[H_r(z,t)] = -\frac{E_o^r}{\eta_o} \cos(\omega t + \beta_1 z) \tag{6.3h}$$

Another important parameter is the time delay of arrival of the reflected field to the transmitter.

If the transmitted fields are reflected off a conductive metallic body R distance away from the transmitter, the time taken for the field to hit that target and come back to the receiver is $\frac{2R}{c}$, where c is the speed of light. Hence the reflected fields of a monostatic radar become:

$$E_r(z,t) = -E_o^i \cos\left(\omega\left(t - \frac{2R}{c}\right) + \beta_1 z\right) a_x \tag{6.3i}$$

$$H_r(z,t) = \frac{E_o^i}{\eta_o} \cos\left(\omega\left(t - \frac{2R}{c}\right) + \beta_1 z\right) a_y \tag{6.3j}$$

When the radar target is moving at a velocity of v, then the radar range $R(t) = R + vt$. R is the radial distance from the source to the target. In this case, the scattered fields are:

$$E_r(z,t) = -E_o^i \cos\left(\omega\left(\left(\frac{1 - \frac{v}{c}}{1 + \frac{v}{c}}\right)\left(t - \frac{R}{c}\right) - \frac{R}{c}\right) + \beta_1 z\right) a_x \tag{6.3k}$$

$$H_r(z,t) = \frac{E_o^i}{\eta_o} \cos\left(\omega\left(\left(\frac{1 - \frac{v}{c}}{1 + \frac{v}{c}}\right)\left(t - \frac{R}{c}\right) - \frac{R}{c}\right) + \beta_1 z\right) a_y \tag{6.3l}$$

The factor, $\left(\frac{1 - \frac{v}{c}}{1 + \frac{v}{c}}\right)$ is called the Doppler scale factor. The interested reader is encouraged to investigate the references for this derivation.

Reference

• Website: https://www.math.colostate.edu/~mueller/graduate_workshop/pdfs_of_talks/ Muller_radarintro.pdf, accessed June 27, 2024.

Application 69: Reflection and Transmission of Obliquely Incident Waves

Let's consider a scenario when a wave hits a boundary at an oblique angle. This scenario is experienced when the source and the boundary are located close to each other. Let's consider a boundary at $y = 0$, and the wave is propagating at an arbitrary direction a_r. Let's also assume that the media are lossless.

I. Oblique incidence with a tangential electric field

In the first case shown in Fig. 6.2, let's consider the oblique incidence with a tangential electric field (i.e. the electric field is parallel to the boundary). The first boundary condition is at $y = 0$ the tangential electric fields should be equal on both sides of the boundary. For the electric field, the tangential direction is the a_x direction.

$$E_o^i + E_o^r = E_o^t \tag{6.4a}$$

The second boundary condition is at $y = 0$ the tangential magnetic fields should be equal. In this case, the tangential direction is the a_z direction.

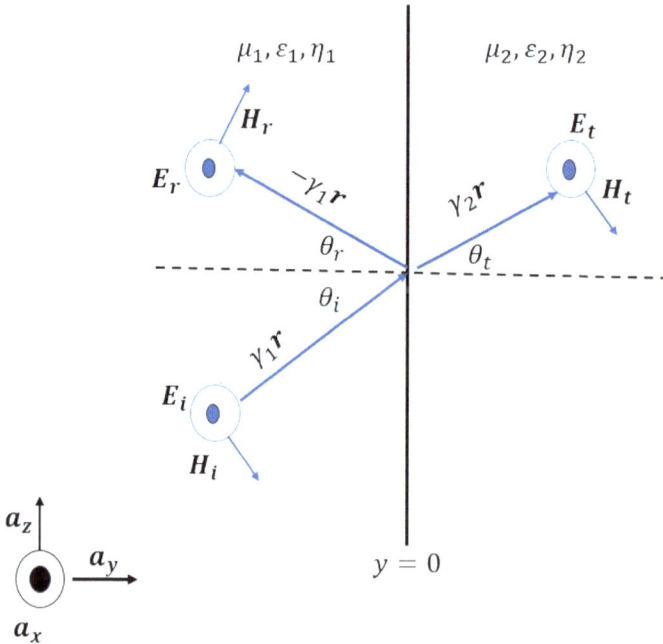

Fig. 6.2 Oblique incidence with tangential electric fields

$$\frac{E_o^i}{\eta_1}(-\cos\theta_i) + \frac{E_o^r}{\eta_1}(\cos\theta_r) = \frac{E_o^t}{\eta_2}(-\cos\theta_t) \tag{6.4b}$$

The above two equations lead to the transmission and reflection coefficients for oblique incidence in the tangential electric field's case as:

$$\tau_E = \frac{E_o^t}{E_o^i} = \frac{\eta_2\cos\theta_i + \eta_2\cos\theta_r}{\eta_2\cos\theta_r + \eta_1\cos\theta_t} \tag{6.4c}$$

$$\Gamma_E = \frac{E_o^r}{E_o^i} = \frac{\eta_2\cos\theta_i - \eta_1\cos\theta_t}{\eta_2\cos\theta_r + \eta_1\cos\theta_t} \tag{6.4d}$$

Given that the incidence and the reflected angles are equations to each other the above two expressions can be simplified to:

$$\tau_E = \frac{E_o^t}{E_o^i} = \frac{2\eta_2\cos\theta_i}{\eta_2\cos\theta_i + \eta_1\cos\theta_t} \tag{6.4e}$$

$$\Gamma_E = \frac{E_o^r}{E_o^i} = \frac{\eta_2\cos\theta_i - \eta_1\cos\theta_t}{\eta_2\cos\theta_i + \eta_1\cos\theta_t} \tag{6.4f}$$

II. Oblique incidence with a tangential magnetic field

let's start by equating the tangential magnetic fields in the a_x direction as shown in Fig. 6.3.

$$H_o^i + H_o^r = H_o^t \tag{6.4g}$$

$$\frac{E_o^i}{\eta_1} + \frac{E_o^r}{\eta_1} = \frac{E_o^t}{\eta_2} \tag{6.4h}$$

And for the tangential electric fields, those will be in the a_z direction.

$$E_o^i\cos\theta_i - E_o^r\cos\theta_r = E_o^t\cos\theta_t \tag{6.4i}$$

$$\tau_E = \frac{E_o^t}{E_o^i} = \frac{\eta_2\cos\theta_i + \eta_2\cos\theta_r}{\eta_1\cos\theta_r + \eta_2\cos\theta_t} \tag{6.4j}$$

$$\Gamma_E = \frac{E_o^r}{E_o^i} = \frac{\eta_1\cos\theta_i - \eta_2\cos\theta_t}{\eta_1\cos\theta_r + \eta_2\cos\theta_t} \tag{6.4k}$$

Once again using the fact that $\theta_i = \theta_r$:

$$\tau_E = \frac{E_o^t}{E_o^i} = \frac{2\eta_2\cos\theta_i}{\eta_2\cos\theta_t + \eta_1\cos\theta_i} \tag{6.4l}$$

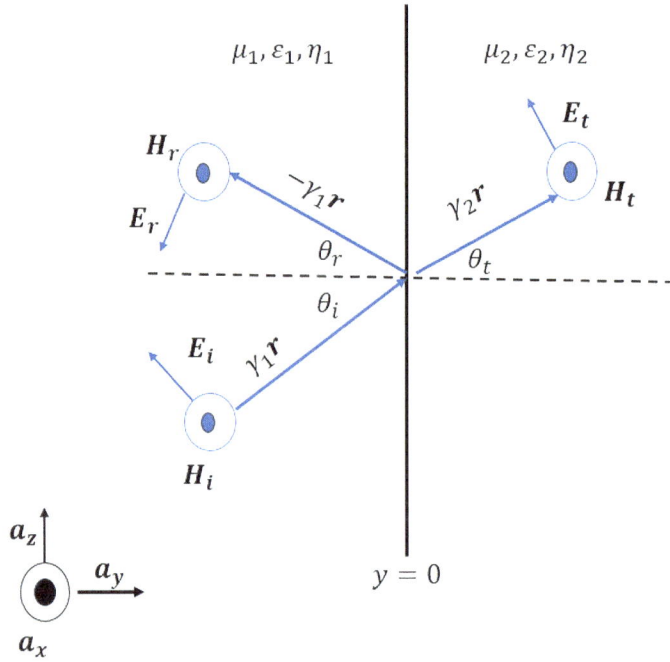

Fig. 6.3 Oblique incidence with tangential magnetic fields

$$\Gamma_E = \frac{E_o^r}{E_o^i} = \frac{\eta_1 \cos\theta_i - \eta_2 \cos\theta_t}{\eta_2 \cos\theta_t + \eta_1 \cos\theta_i} \tag{6.4m}$$

Reference

- Website: https://eng.libretexts.org/Bookshelves/Electrical_Engineering/Electro-Opt ics/Electromagnetic_Field_Theory%3A_A_Problem_Solving_Approach_(Zahn)/07% 3A_Electrodynamics-fields_and_Waves/7.09%3A_Oblique_Incidence_Onto_a_Dielec tric#:~:text=Figure%207%2D18%20A%20uniform,is%20given%20by%20Snell's% 20law. Accessed June 25, 2024.

Application 70: Poynting Theorem and the Electromagnetic Power

The Poynting theorem indicates the net flux coming into a volume v through a surface s is:

$$P_{net,in} = -\oint_s E \times H ds \tag{6.5}$$

A reader interested in the derivation of this theorem is referred to the website of Virginia Polytechnique University.

$$P_{net,in} = -\oint_s E \times H ds = P_{Ohmic} + P_{electric} + P_{magnetetic} \tag{6.5a}$$

$$P_{Ohmic} = \int_v E.J dv = \int_v \sigma |E|^2 dv; \because J = \sigma E \tag{6.5b}$$

$$P_{electric} = \frac{1}{2}\frac{\partial}{\partial t}\int_v E.D dv = \frac{1}{2}\frac{\partial}{\partial t}(\int_v \varepsilon |E|^2 dv); \because D = \varepsilon E \tag{6.5c}$$

$$P_{magnetic} = \frac{1}{2}\frac{\partial}{\partial t}\int_v H.B dv = \frac{1}{2}\frac{\partial}{\partial t}(\int_v \mu |H|^2 dv); \because B = \mu H \tag{6.5d}$$

For current and voltage for time-harmonic signals:

$$P_{avg} = v_{rms}i_{rms} = \frac{1}{2}vi \tag{6.5e}$$

For complex signals:

$$P_{avg} = \frac{1}{2}Re\{vi^*\} \tag{6.5f}$$

In Eq. 6.5f, * indicates the complex conjugate. Using the same analogy for electric and magnetic fields.

$$P_{avg,density} = \frac{1}{2}Re\{E \times H^*\} \tag{6.5g}$$

Electric and magnetic fields are in orthogonal planes; hence the cross product and the power propagation will be orthogonal to both of those. And this defines the power density or the power through a unit surface area. An easy way to remember this is by paying attention to the units of both electric and magnetic field intensities. Both quantities have per meter in their denominator hence giving per square meter in the power quantity, making it a power per unit surface area.

Average Power Density in Free Space

Considering only the forward traveling waves, and since the attenuation in free space is zero, the time-harmonic electric and magnetic fields become:

$$E(z,t) = E_o^+ e^{-\beta z} e^{j\omega t} \mathbf{a_x} \tag{6.5h}$$

$$H(z,t) = \frac{E_o^+}{\eta_o} e^{-\beta z} e^{j\omega t} \mathbf{a_y} \tag{6.5i}$$

$$P_{avg.density} = \frac{1}{2} Re\left\{ E_o^+ e^{-\beta z} e^{j\omega t} \mathbf{a_x} \times \frac{E_o^+}{\eta_o} e^{+\beta z} e^{-j\omega t} \mathbf{a_y} \right\} \tag{6.5j}$$

$$P_{avg.density} = \frac{1}{2} Re\left\{ \frac{(E_o^+)^2}{\eta_o} \mathbf{a_z} \right\} \tag{6.5k}$$

$$P_{avg.density} = \frac{1}{2} \frac{(E_o^+)^2}{\eta_o} \mathbf{a_z} \tag{6.5l}$$

Average Power Density in Lossy Media

$$E(z,t) = E_o^+ e^{-\gamma z} e^{j\omega t} \mathbf{a_x} \tag{6.5m}$$

$$H(z,t) = \frac{E_o^+}{\eta} e^{-\gamma z} e^{j\omega t} \mathbf{a_y} \tag{6.5n}$$

Since $\gamma = \alpha + j\beta$; $e^{-\gamma z} = e^{-\alpha z} e^{-j\beta z}$, hence

$$E(z,t) = E_o^+ e^{-\alpha z} e^{-j\beta z} e^{j\omega t} \mathbf{a_x} \tag{6.5o}$$

$$H(z,t) = \frac{E_o^+}{\eta} e^{-\alpha z} e^{-j\beta z} e^{j\omega t} \mathbf{a_y} \tag{6.5p}$$

In a lossy media, the impedance η is complex, hence can be written as: $|\eta| e^{j\theta}$. The power density expression becomes:

$$P_{avg.density} = \frac{1}{2} Re\left\{ E_o^+ e^{-\alpha z} e^{-j\beta z} e^{j\omega t} \mathbf{a_x} \times \frac{E_o^+}{|\eta|} e^{-\alpha z} e^{+\beta z} e^{+j\theta} e^{-j\omega t} \mathbf{a_y} \right\} \tag{6.5q}$$

$$P_{avg.density} = \frac{1}{2} Re\left\{ \frac{(E_o^+)^2}{|\eta|} e^{-2\alpha z} e^{j\theta} \mathbf{a_z} \right\} \tag{6.5r}$$

$$P_{avg,density} = \frac{1}{2}\frac{(E_o^+)^2}{|\eta|}e^{-2\alpha z}\cos(\theta)a_z \qquad (6.5s)$$

Application 71: Radiation Fields of Antennas

Let's consider a dipole antenna along the z-axis as shown in Fig. 6.4.

For the derivation of the radiation fields, we use a purely mathematical quantity called the magnetic vector potential A. The relationship of the magnetic vector potential to the magnetic flux density is as follows:

From the Gauss's law for magnetic fields or the solenoidal law says:

$$\nabla.B = 0 \qquad (6.6)$$

We choose A such that, $\nabla.(\nabla \times A) = 0$. Hence, $B = \nabla \times A$.

For the dipole above, $R = (x - 0)a_x + (y - 0)a_y + (z - 0)a_z$.

$$R = xa_x + ya_y + za_z \qquad (6.6a)$$

$$|R| = \sqrt{x^2 + y^2 + z^2} \qquad (6.6b)$$

Fig. 6.4 A dipole antenna along the z axis

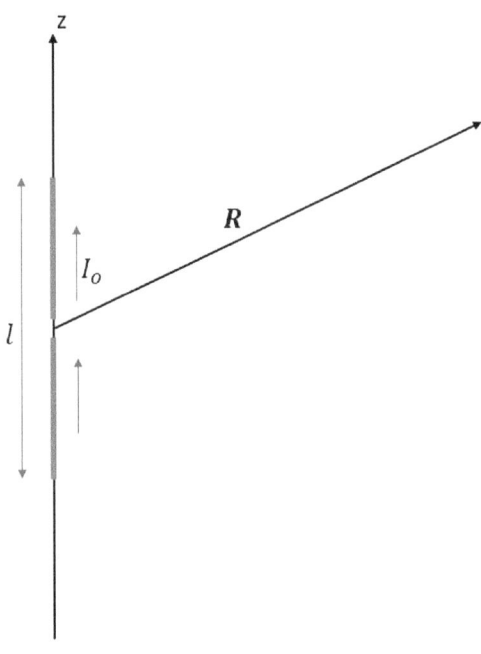

The magnetic vector potential created by a thin current-carrying conductor is:

$$A = \frac{\mu_0}{4\pi} \int_l I_e(x', y', z') \frac{e^{-jkR}}{R} .dl \tag{6.6c}$$

$$A = \frac{\mu_0}{4\pi R} e^{-jkR} \int_{-l/2}^{l/2} I_o a_z .dz = \frac{\mu_0 I_o l}{4\pi R} e^{-jkR} a_z \tag{6.6d}$$

From the above equation, we can see that only the component exists. Now let's transform the coordinate system to spherical since the antenna radiation is omnidirectional. Also in the far field, $R \approx r$.

$$\begin{bmatrix} A_r \\ A_\theta \\ A_\phi \end{bmatrix} = \begin{bmatrix} sin\theta cos\phi & sin\theta sin\phi & cos\theta \\ cos\theta sin\phi & cos\theta sin\phi & -sin\theta \\ -sin\phi & cos\phi & 0 \end{bmatrix} \begin{bmatrix} A_x \\ A_y \\ A_z \end{bmatrix} \tag{6.6e}$$

$$A_r = cos\theta A_z \tag{6.6f}$$

$$A_\theta = -sin\theta A_z \tag{6.6g}$$

$$A_\phi = 0 \tag{6.6h}$$

From the definition of the magnetic vector potential.

$$H = \frac{1}{\mu_0}(\nabla \times A) \tag{6.6i}$$

Given that $A_\phi = 0$ and the other two components do not have ϕ dependencies, the curl expression in spherical coordinates reduces to:

$$H = \frac{1}{\mu_0 r}\left(\frac{\partial(rA_\theta)}{\partial r} - \frac{\partial(A_r)}{\partial\theta}\right)a_\phi \tag{6.6j}$$

$$H = \frac{jkI_o l}{4\pi R} e^{-jkR}\left(1 + \frac{1}{jkr}\right)sin\theta a_\phi \tag{6.6k}$$

And we can derive the electric field expressions from Ampere's Law in point form.

$$\nabla \times H = \frac{\partial D}{\partial t} + J \tag{6.6l}$$

We assume that the region is currently free, and the fields are time-harmonic. Hence:

$$\nabla \times H = \frac{\partial \varepsilon e^{j\omega t} E_s}{\partial t} \tag{6.6m}$$

$$E = \frac{1}{j\omega\varepsilon}(\nabla \times H) \tag{6.6n}$$

Now, let's evaluate the components of the electric field separately in each direction. Given that the magnetic field intensity is in the a_ϕ direction, the non-zero electric field components will be in the a_r and a_θ directions.

$$E_r = \frac{1}{j\omega\varepsilon r \sin\theta}\left(\frac{\partial}{\partial\theta}(H_\phi \sin\theta)\right) \tag{6.6o}$$

Evaluation of this would yield:

$$E_r = \frac{k}{\omega\varepsilon}\frac{I_o l}{2\pi r^2}e^{-jkr}\left(1 + \frac{1}{jkr}\right)\cos\theta \tag{6.6p}$$

The $\frac{k}{\omega\varepsilon}$ term reduces to $\frac{1}{\sqrt{\mu_o\varepsilon_o}}$. Therefore, the final expression for the radial electric field intensity would be:

$$E_r = \frac{1}{\sqrt{\mu_o\varepsilon_o}}\frac{I_o l}{2\pi r^2}e^{-jkr}\left(1 + \frac{1}{jkr}\right)\cos\theta \tag{6.6q}$$

Now, let's consider the electric field intensity in the a_θ direction.

$$E_\theta = \frac{1}{j\omega\varepsilon r}\left(-\frac{\partial}{\partial r}(rH_\phi)\right) \tag{6.6r}$$

$$E_\theta = \frac{k}{\omega\varepsilon}\frac{I_o l}{4\pi r}\left(-\frac{\partial}{\partial r}\left(e^{-jkr}\left(1 + \frac{1}{jkr}\right)\right)\right)\sin\theta \tag{6.6s}$$

The term $\frac{k}{\omega\varepsilon}$ is the free space impedance η_o. Hence the electric field intensity in the a_θ direction is:

$$E_\theta = \eta_o\frac{I_o l}{4\pi r}jke^{-jkr}\left(1 + \frac{1}{jkr} - \frac{1}{(kr)^2}\right)\sin\theta \tag{6.6t}$$

The full electric field intensity expression becomes:

$$E = \frac{1}{\sqrt{\mu_o\varepsilon_o}}\frac{I_o l}{2\pi r^2}e^{-jkr}\left(1 + \frac{1}{jkr}\right)\cos\theta a_r + \eta_o\frac{I_o l}{4\pi r}jke^{-jkr}\left(1 + \frac{1}{jkr} - \frac{1}{(kr)^2}\right)\sin\theta a_\theta \tag{6.6u}$$

Application 72: Characteristic Impedance of a Transmission Line

Transmission lines are used to carry high-frequency voltage and current signals. A typical transmission line terminates at an antenna where that voltage and current signal induces time-varying electric and magnetic fields on the antenna structure initiating the electromagnetic wave propagation. At the receiving end, an electromagnetic wave induces time-varying currents and voltages on the antenna, which are then carried to the high-frequency circuitry by the transmission lines.

The transmission lines are pairs of wires. The coaxial cables, twisted pairs, and etched copper transmission lines on printed circuit boards are all considered transmission lines. Figure 6.5 shows two printed circuit board type transmission lines created on Ansys HFSS™. When dealing with transmission lines one thing to consider is the self-impedance of the transmission lines. This impedance is called the characteristic impedance. The characteristic impedance per unit length can be modeled by a lumped element circuit model as given in Fig. 6.6.

Let's consider the coaxial transmission line. Previously we derived the capacitance ($C\prime$) and the inductance ($L\prime$) per unit length of the coaxial cable. Note that the capacitance per unit length is a function of the permittivity of the low-loss dielectric material and the inductance is a function of the permeability of the low-loss dielectric.

$$C' = \frac{2\pi\varepsilon}{\ln\left(\frac{b}{a}\right)} \tag{6.7a}$$

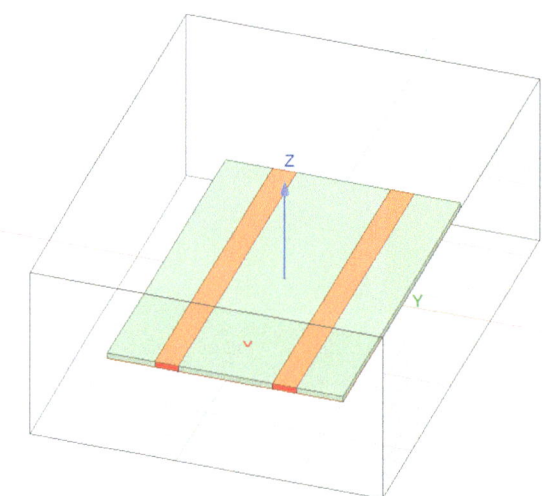

Fig. 6.5 Printed circuit board transmission lines created on Ansys HFSS

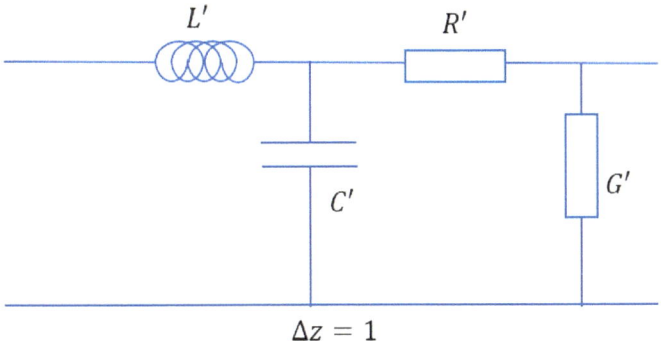

Fig. 6.6 Lumped element circuit model for a unit length of a transmission line

$$L' = \frac{\mu}{2\pi} \ln\left(\frac{b}{a}\right) \tag{6.7b}$$

Let's derive the conductance and the resistance per unit length of this transmission line. As expected, the conductance per unit length will be a function of the conductivity of the low-loss dielectric material between the inner current-carrying conductor and the outer copper mesh (the outer conductor). Let's consider the conductivity of the low-loss dielectric as σ_d. The conductance occurs due to the leakage of current between the two conductors in the $\boldsymbol{a}_\rho$ direction.

$$J_\rho \boldsymbol{a}_\rho = \sigma_d E_\rho \boldsymbol{a}_\rho \tag{6.7c}$$

$$E_\rho = \frac{J_\rho}{\sigma_d} = \frac{I}{2\pi \rho h \sigma_d} \tag{6.7d}$$

The electric potential difference between the two conductors Φ :

$$\Phi = -\int_b^a \frac{I}{2\pi \rho h \sigma_d} . d\rho \tag{6.7e}$$

$$\Phi = \frac{I}{2\pi h \sigma_d} \ln\left(\frac{b}{a}\right) \tag{6.7f}$$

The conductance per a length (height) h of a coaxial cable, $G = I/\Phi$.

$$G = \frac{2\pi h \sigma_d}{\ln\left(\frac{b}{a}\right)} \tag{6.7g}$$

And the conductance per unit length of the coaxial cable $G\prime$:

$$G' = \frac{2\pi\sigma_d}{\ln\left(\frac{b}{a}\right)} \tag{6.7h}$$

Now, let's calculate the resistance per unit length of the coaxial cable. the resistance of any conductor with length L, and conductivity σ_c and cross-sectional area of current flow A, is given by:

$$R = \frac{1}{\sigma_c}\left(\frac{L}{A}\right) \tag{6.7i}$$

For the coaxial cable, both the inner and outer conductors are made of copper. Resistance per unit length of the inner and outer conductors are:

$$R'_{ic} = \frac{1}{\sigma_c}\left(\frac{1}{2\pi a\delta_c}\right) \tag{6.7j}$$

$$R'_{oc} = \frac{1}{\sigma_c}\left(\frac{1}{2\pi b\delta_c}\right) \tag{6.7k}$$

δ_c is the skin depth of the conductor. Which is the reciprocal of the attenuation constant. Since both conductors are made with copper, the attenuation constant would be the same for both conductors, hence the same skin depth.

$$R' = R'_{ic} + R'_{oc} \tag{6.7l}$$

$$R' = \frac{1}{2\pi\sigma_c\delta_c}\left(\frac{1}{a} + \frac{1}{b}\right) \tag{6.7m}$$

For electric conductors the attenuation constant $\alpha = \sqrt{\frac{\omega\mu\sigma}{2}}$, where μ in this case the permeability of the conductor. For a good conductor such as copper, the permeability is the same as the permeability of free space(μ_o). Therefore,

$$R' = \frac{1}{2\pi}\left(\frac{1}{a} + \frac{1}{b}\right)\sqrt{\frac{\omega\mu_o}{2\sigma_c}} \tag{6.7n}$$

Let's consider the voltage difference between $z = z$ and $z = z + \Delta z$ locations.

$$v(z, t) - v(z + \Delta z, t) = R'\Delta z i(z, t) + L'\Delta z\frac{di(z, t)}{dt} \tag{6.7o}$$

Now let's consider, $\displaystyle\lim_{\Delta z \to 0} \frac{(v(z,t) - v(z+\Delta z,t))}{\Delta z} = -\frac{\partial v(z,t)}{\partial z}$

$$-\frac{\partial v(z, t)}{\partial z} = R'i(z, t) + L'\frac{di(z, t)}{dt} \tag{6.7p}$$

Similarly, by taking the parallel combination of the conductance and the capacitance:

$$-\frac{\partial i(z, t)}{\partial z} = G'v(z, t) + C'\frac{dv(z, t)}{dt} \tag{6.7q}$$

Now, let's assume that both the voltage and current are time-harmonic fields. This is a valid assumption given that all human-made signals used in transmission are sinusoidal.

$$v(z, t) = Re\left\{V_s(z)e^{j\omega t}\right\} \tag{6.7r}$$

$$i(z, t) = Re\left\{I_s(z)e^{j\omega t}\right\} \tag{6.7s}$$

$$\frac{\partial v(z, t)}{\partial z} = \frac{dV_s(z)}{dz} = -(R' + j\omega L')I_s(z) \tag{6.7t}$$

$$\frac{\partial i(z, t)}{\partial z} = \frac{dI_s(z)}{dz} = -(G' + j\omega C')V_s(z) \tag{6.7u}$$

Now, let's take the second derivative of one of the above equations:

$$\frac{d}{dz}\left(\frac{dV_s(z)}{dz}\right) = -(R' + j\omega L')\frac{dI_s(z)}{dz} \tag{6.7v}$$

$$\frac{d^2 V_s(z)}{dz^2} = (R' + j\omega L')(G' + j\omega C')V_s(z) \tag{6.7w}$$

$$\frac{d^2 V_s(z)}{dz^2} - \gamma^2 V_s(z) = 0 \tag{6.7x}$$

$$V_s(z) = V_o^+ e^{-\gamma z} + V_o^- e^{+\gamma z} \tag{6.7y}$$

Similarly for the current:

$$I_s(z) = I_o^+ e^{-\gamma z} + I_o^- e^{+\gamma z} \tag{6.7z}$$

The propagation constant of a transmission line is:

$$\gamma = \sqrt{(R' + j\omega L')(G' + j\omega C')} \tag{6.7aa}$$

$$\frac{dV_s(z)}{dz} = -\gamma\left(V_o^+ e^{-\gamma z} - V_o^- e^{+\gamma z}\right) = -(R' + j\omega L')I_s(z) \tag{6.7ab}$$

$$I_s(z) = \frac{\gamma}{(R' + j\omega L')}\left(V_o^+ e^{-\gamma z} - V_o^- e^{+\gamma z}\right) \tag{6.7ac}$$

The characteristic impedance Z_o:

$$Z_0 = \frac{V_o^+}{I_o^+} = -\frac{V_o^-}{I_o^-} = \sqrt{\frac{R' + j\omega L'}{G' + j\omega C'}} \qquad (6.7ad)$$

And

$$I_s(z) = \frac{1}{Z_0}\left(V_o^+ e^{-\gamma z} - V_o^- e^{+\gamma z}\right) \qquad (6.7ae)$$

Application 73: Attenuation and Phase Constants of a Transmission Line

The propagation constant of a transmission line is defined by γ. Like the propagating waves, the real part of γ is called the attenuation constant and the imaginary part of γ is called the phase constant.

$$\gamma = \sqrt{(R' + j\omega L')(G' + j\omega C')} \qquad (6.8a)$$

$$\gamma = \alpha + j\beta \qquad (6.8b)$$

Lossless Transmission Lines

When the transmission line is lossless, the resistance and the conductance per unit length become zero. Hence the above equations are reduced to:

$$Z_0 = \sqrt{\frac{L'}{C'}} = \sqrt{\frac{\frac{\mu}{2\pi}\ln\left(\frac{b}{a}\right)}{\frac{2\pi\varepsilon}{\ln\left(\frac{b}{a}\right)}}} = \sqrt{\frac{\mu_o}{\varepsilon_o}}\frac{1}{2\pi}\ln\left(\frac{b}{a}\right)\sqrt{\frac{\mu_r}{\varepsilon_r}} \qquad (6.8c)$$

$\sqrt{\frac{\mu_o}{\varepsilon_o}} = \eta_o$ the free space impedance. Hence,

$$Z_0 = \frac{\eta_o}{2\pi}\ln\left(\frac{b}{a}\right)\sqrt{\frac{\mu_r}{\varepsilon_r}} \qquad (6.8d)$$

$$\alpha = 0$$

$$\beta = \omega\sqrt{L'C'} = \omega\sqrt{\frac{\mu}{2\pi}\ln\left(\frac{b}{a}\right)\cdot\frac{2\pi\varepsilon}{\ln\left(\frac{b}{a}\right)}} = \omega\sqrt{\mu_o\varepsilon_o}\cdot\sqrt{\mu_r\varepsilon_r} \qquad (6.8e)$$

$$\sqrt{\mu_0 \varepsilon_0} = \frac{1}{c} \qquad (6.8\text{f})$$

Hence for the loss-less case, the phase constant becomes:

$$\beta = \frac{\omega}{c}\sqrt{\mu_r \varepsilon_r} \qquad (6.8\text{g})$$

Low-Loss Transmission Lines

Now, let's consider a low-loss transmission line. For a low-loss transmission line, the $R\prime$ and $G\prime$ values are small but not zero. Let us start with the propagation constant:

$$\gamma = \sqrt{(R' + j\omega L')(G' + j\omega C')} = \sqrt{(j\omega L')(j\omega C')}\sqrt{\left(1 + \frac{R'}{j\omega L'}\right)\left(1 + \frac{G'}{j\omega C'}\right)} \qquad (6.8\text{h})$$

$$\gamma = \sqrt{(j\omega L')(j\omega C')}\sqrt{1 - j\left(\frac{R'}{\omega L'} + \frac{G'}{\omega C'}\right) - \frac{R'G'}{\omega^2 L'C'}} \qquad (6.8\text{i})$$

Given that both $R\prime$ and $G\prime$ are small, the quantity $\frac{R'G'}{\omega^2 L'C'}$ will be nearly zero. Hence, we can neglect that and derive the following approximation for γ.

$$\gamma \approx \sqrt{(j\omega L')(j\omega C')}\sqrt{1 - j\left(\frac{R'}{\omega L'} + \frac{G'}{\omega C'}\right)} \qquad (6.8\text{j})$$

Now, using the Taylor series expansion for small a:

$$(1 + a)^n = 1 + na \qquad (6.8\text{k})$$

$$\left(1 - j\left(\frac{R'}{\omega L'} + \frac{G'}{\omega C'}\right)\right)^{1/2} = 1 - j\frac{1}{2}\left(\frac{R'}{\omega L'} + \frac{G'}{\omega C'}\right) \qquad (6.8\text{l})$$

And the propagation constant becomes:

$$\gamma \approx j\omega\sqrt{L'C'}\left(1 - j\frac{1}{2}\left(\frac{R'}{\omega L'} + \frac{G'}{\omega C'}\right)\right) \qquad (6.8\text{m})$$

The attenuation constant of a low-loss transmission is:

$$\alpha = Re\{\gamma\} = \frac{1}{2}\sqrt{L'C'}\left(\frac{R'}{L'} + \frac{G'}{C'}\right) \qquad (6.8\text{n})$$

And the phase constant β for the low-loss case is:

$$\beta = Im\{\gamma\} = \omega\sqrt{L'C'} \qquad (6.8\text{o})$$

Distortion Less Transmission Lines

For a transmission line to be distortion less the attenuation constant and the characteristic impedance both should be independent of frequency. The condition for that is:

$$\frac{R'}{L'} = \frac{G'}{C'} \rightarrow G' = \frac{R'C'}{L'} \qquad (6.8\text{p})$$

$$\gamma = \sqrt{(R' + j\omega L')(G' + j\omega C')} = \sqrt{(R' + j\omega L')\left(\frac{R'C'}{L'} + j\omega C'\right)} \qquad (6.8\text{q})$$

$$\gamma = \sqrt{(R' + j\omega L')\left(\frac{R'C' + j\omega L'C'}{L'}\right)} = (R' + j\omega L')\sqrt{\frac{C'}{L'}} \qquad (6.8\text{r})$$

$$\alpha = Re\{\gamma\} = R'\sqrt{\frac{C'}{L'}} \qquad (6.8\text{s})$$

$$\beta = Im\{\gamma\} = \omega\sqrt{L'C'} \qquad (6.8\text{t})$$

Application 74: Transmission Line Reflection Coefficient and Input Impedance

A typical transmission line is terminated by a load such as an antenna. Let's consider the case where a transmission line with the characteristic impedance Z_o is terminated by a load (antenna) with a load impedance of Z_L. For the easiness of analysis, let's assume that the location of the load is at $z = 0$.

$$Z_L = \frac{V_s(z = 0)}{I_s(z = 0)} = \frac{V_o^+ + V_o^-}{I_o^+ + I_o^-} = Z_o\left(\frac{V_o^+ + V_o^-}{V_o^+ - V_o^-}\right) \qquad (6.9\text{a})$$

The voltage reflection coefficient is given by:

$$\Gamma = \frac{V_o^-}{V_o^+} = \frac{Z_L - Z_o}{Z_L + Z_o} \qquad (6.9\text{b})$$

Now, to find the input impedance let's consider a distance of $-l$ from the load.

$$Z_{in} = \frac{V_s(z = -l)}{I_s(z = -l)} = Z_o \left(\frac{V_o^+ e^{+\gamma l} + V_o^- e^{-\gamma l}}{V_o^+ e^{+\gamma l} - V_o^- e^{-\gamma l}} \right) \tag{6.9c}$$

From the reflection coefficient relationship:

$$V_o^- = \left(\frac{Z_L - Z_o}{Z_L + Z_o} \right) V_o^+ \tag{6.9d}$$

Also, let's assume that the transmission line is lossless, hence the attenuation constant $\alpha = 0$.

$$Z_{in} = Z_o \left\{ \frac{(Z_L + Z_o)e^{+j\beta l} + (Z_L - Z_o)e^{-j\beta l}}{(Z_L + Z_o)e^{+j\beta l} - (Z_L - Z_o)e^{-j\beta l}} \right\} \tag{6.9e}$$

$$Z_{in} = Z_o \left\{ \frac{Z_L(e^{+j\beta l} + e^{-j\beta l}) + Z_o(e^{+j\beta l} - e^{-j\beta l})}{Z_o(e^{+j\beta l} + e^{-j\beta l}) + Z_L(e^{+j\beta l} - e^{-j\beta l})} \right\} \tag{6.9f}$$

From the Euler's expansion:

$$e^{+j\beta l} + e^{-j\beta l} = 2\cos(\beta l) \tag{6.9g}$$

$$e^{+j\beta l} - e^{-j\beta l} = 2j\sin(\beta l) \tag{6.9h}$$

Hence:

$$Z_{in} = Z_o \left\{ \frac{Z_L\cos(\beta l) + jZ_o\sin(\beta l)}{Z_o\cos(\beta l) + jZ_L\sin(\beta l)} \right\} \tag{6.9i}$$

Dividing both the numerator and the denominator by the cosine function:

$$Z_{in} = Z_o \left\{ \frac{Z_L + jZ_o\tan(\beta l)}{Z_o + jZ_L\tan(\beta l)} \right\} \tag{6.9j}$$

The input impedance of an antenna plays an important role in impedance matching. The interested reader can refer to the quarter-wave transformer matching and stub-matching techniques which use the above formulas to match the transmission line impedance to the load impedance.

Application 75: Wireless Communications

Wireless communication is the pinnacle application of Maxwell's equations. The invention of wireless communications revolutionized the world. Starting from wireless telephone communications, internet, and satellite communications, wireless communications connected the world. All forms of wireless communications use electromagnetic wave theory

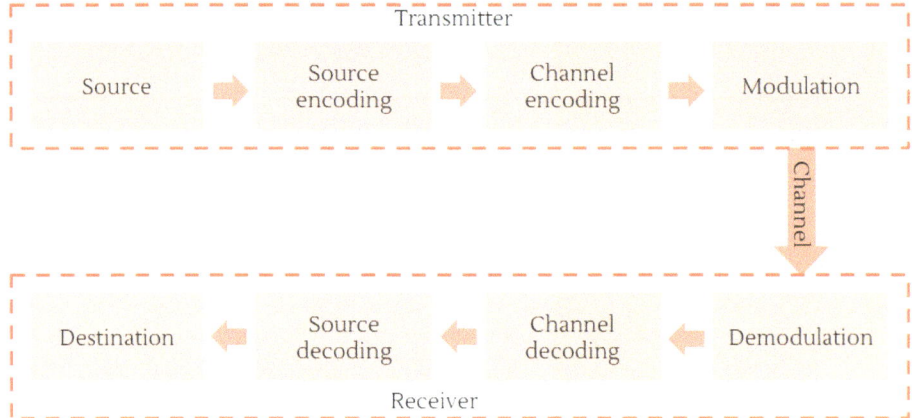

Fig. 6.7 Processes of a point-to-point digital communication system

for signal propagation. Although the signal processing techniques change from one wireless communication system to another, the wave propagation theory between the transmitter antenna to the receiver antenna does not change. For demonstration purposes here we show the main processes of a digital communication system in Fig. 6.7.

The analog signal is digitized using the Shannon-Nyquist sampling theorem, quantized, and encoded. This is known as source encoding. More bits are added at the input of the channel for error correction and then converted back to an analog signal before transmission. The flow chart below shows the steps involved before transmission and after reception. Maxwell's equations play a role in the propagation between the transmitter and receiver antennas.

A transmission line brings the signal to the transmitter antenna. This signal is a high-frequency time-varying signal. This high-frequency time-varying voltage signal creates a high-frequency time-varying current signal within the transmission line. When this signal is delivered to the load antenna, the time-varying current generates a time-varying magnetic field. This time-varying magnetic field generates a time-varying electric field. This process repeats and the time-varying electro-magnetic wave transmits electromagnetic power according to the Poynting theorem.

At the receiving end, the electromagnetic wave induces currents and voltages on the receiving antenna which is then connected to another transmission line that carries the power in the electric circuitry.

Application 76: Refractive Index and Electric Polarization

In the case above, we saw that the homogeneous wave equation became:

$$\frac{\partial^2 E}{\partial z^2} = \frac{1}{c^2}\frac{\partial}{\partial t}\left(\frac{\partial E}{\partial t}\right) \tag{6.10a}$$

Let's consider wave propagation inside a medium other than free space. If the speed of propagation of an electromagnetic wave in that medium is v in that case, the homogeneous wave equation becomes:

$$\frac{\partial^2 E}{\partial z^2} = \frac{1}{v^2}\frac{\partial}{\partial t}\left(\frac{\partial E}{\partial t}\right) \tag{6.10b}$$

$$v^2 = \frac{1}{\mu_r \varepsilon_r \mu_o \varepsilon_o} \tag{6.10c}$$

The refractive index is defined as:

$$n = \frac{c}{v} = \sqrt{\frac{\mu_r \varepsilon_r \mu_o \varepsilon_o}{\mu_o \varepsilon_o}} = \sqrt{\mu_r \varepsilon_r} \tag{6.10d}$$

For dielectric materials, the relative permeability is 1, but the relative permittivity is greater than 1. In that case, the

$$n = \sqrt{\varepsilon_r} \tag{6.10e}$$

And:

$$n^2 = \varepsilon_r = 1 + \chi_e \tag{6.10f}$$

χ_e is the electric susceptibility—the ability of a material to become instantaneously polarized. The amount of polarization for a linear isotropic medium is given by the Eq. 6.10g. And the relationship between the electric polarization P, electric field intensity E and the electric flux density D is given in Eq. 6.10h. And the polarizability is $\alpha = \chi_e \varepsilon_o$.

$$P = \chi_e \varepsilon_o E \tag{6.10g}$$

$$D = \varepsilon_o E + P \tag{6.10h}$$

The relationship given in Eq. 6.10 can be generalized into a linear-time invariant system as shown in Eq. 6.10i. In Eq. 6.10i, r is the position vector and t is time.

$$P(r, t) = \chi_e(r, t) * \varepsilon_o E(r, t) \tag{6.10i}$$

The Fourier transform of Eq. 6.10i gives the following relationship in Eq. 6.10j.

$$P(r, \omega) = \varepsilon_o \chi_e(r, \omega)E(r, \omega) \tag{6.10j}$$

The frequency domain relationship of Eq. 6.10h, for a linear time invariant system is:

$$D(r, \omega) = \varepsilon_o E(r, \omega) + P(r, \omega) \qquad (6.10\text{k})$$

The Eq. 6.10j indicates that the electric susceptibility, electric polarization, and electric field intensity are functions of the frequency, and location. Hence the electric flux density is also a function of position and frequency. For anisotropic media, the electric susceptibility becomes a tensor.

Reference

- Website: http://www1.udel.edu/chem/sneal/sln_tchng/CHEM620/CHEM620/Chi_3._Light-Matter_Interactions_P.html accessed June 25, 2024.

Application 77: Friss Transmission Equation

Friss transmission equation gives a relationship between the transmitted and received power of two antennas. Let's start with the isotropic radiator with a transmission power of P_t. For an isotropic radiator, the power density at a point R distance away would be:

$$P_{t,density} = \frac{P_t}{4\pi R^2} \qquad (6.11\text{a})$$

Let's consider a radiator with where e_t is the transmission efficiency. For an antenna that is fully impedance-matched to the radiation efficiency, the transmission efficiency will be one. Also, let's consider the directivity of the antenna in the direction of the receiver is D_t. Figure 6.8 shows two line-of-sight antennas with different transmission and reception efficiencies.

For a directional, low-loss antenna the received power density becomes:

$$P_{r,density} = e_t D_t \frac{P_t}{4\pi R^2} \qquad (6.11\text{b})$$

The effective aperture of the receiver antenna is:

$$A_{r,eff} = e_r D_r \left(\frac{\lambda^2}{4\pi} \right) \qquad (6.11\text{c})$$

The total received power is:

$$P_r = A_{r,eff} . P_{r,density} \qquad (6.11\text{d})$$

$$\frac{P_r}{P_t} = e_t D_t . e_r D_r \left(\frac{\lambda}{4\pi R} \right)^2 \qquad (6.11\text{e})$$

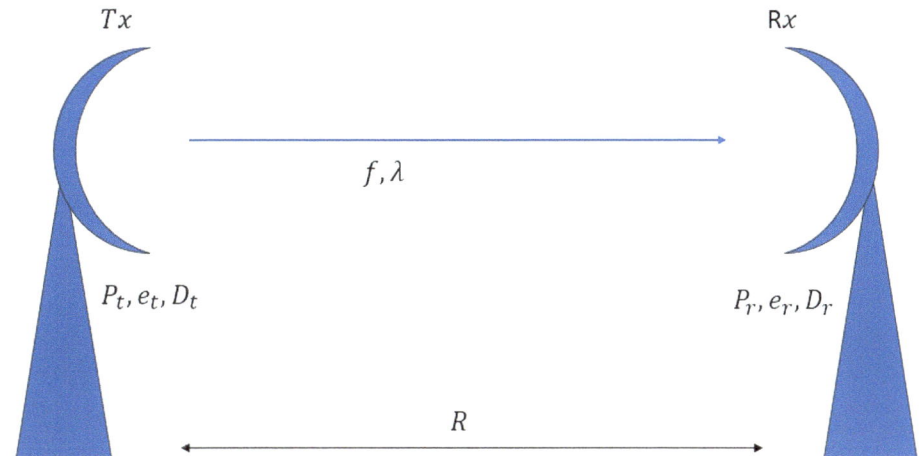

Fig. 6.8 Two non-isotropic radiators focused line-of-sight

$e_t D_t = G_t$ and $e_r D_r = G_r$. Hence:

$$\frac{P_r}{P_t} = G_t G_r \left(\frac{\lambda}{4\pi R} \right)^2 \qquad (6.11f)$$

Application 78: Radar Range Equation

The radar range equation or the radar equation gives the received power of the radar. In this case, remember that the transmission and the receiving both are done using the same antenna as shown in Fig. 6.9.

In a typical radar system, the transmitter sends a radar pulse and waits for the receiving pulse to arrive. The distance is determined by the equation we discussed above, and the received power is determined by the radar equation. Let's start with a directional transmitter with a gain of G_t.

$$P_{t,density} = \frac{P_t G_t}{4\pi R^2} \qquad (6.12a)$$

We are assuming that the target is at a distance of R. If the radar cross section is σ_R, the power received by the radar target is:

$$P_r = \frac{P_t G_t \sigma_R}{4\pi R^2} \qquad (6.12b)$$

Now, this radar target reflects power. If the power is reflected isotopically the received power density of the antenna is:

Fig. 6.9 The radar transmission and reception antenna and the radar cross section

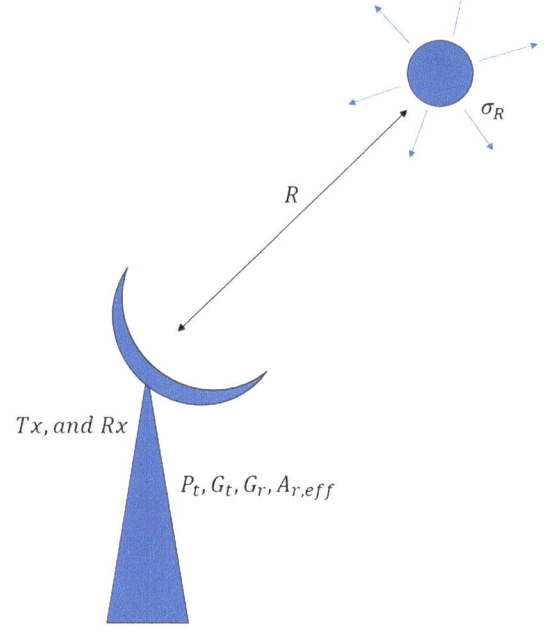

$$P_{r,density} = \frac{P_t G_t}{4\pi R^2} \cdot \frac{\sigma_R}{4\pi R^2} \tag{6.12c}$$

If the receiving gain of the antenna is G_r and the effective aperture is $\frac{\lambda^2}{4\pi}$, the total received power is:

$$P_r = G_r P_{r,density} \frac{\lambda^2}{4\pi} \tag{6.12d}$$

$$P_r = G_t G_r \frac{\sigma_R \lambda^2 P_t}{(4\pi)^3 R^4} \tag{6.12e}$$

Reference

- Website: https://www.ll.mit.edu/sites/default/files/outreach/doc/2018-07/lecture%202. pdf accessed June 25, 2024.

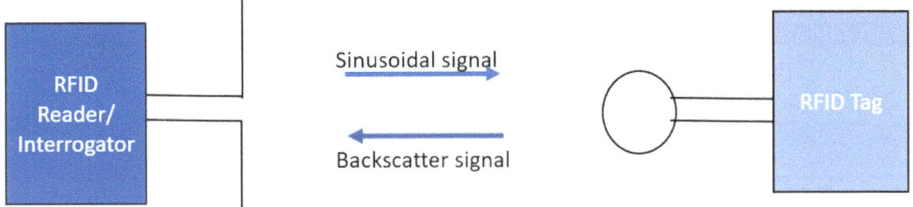

Fig.6.10 The components of a radar telemetry system

Application 79: Radio Telemetry

Radio telemetry is also a radio frequency identification technique. But instead of using near-field inductive coupling, this technique uses backscatter modulation as shown in Fig. 6.10. This is also called far-field electromagnetic coupling. In this case, once the reader or the interrogator is sending signals continuously. Unlike near-field coupling in far-field coupling, there is no magnetic linkage. Instead, once the reader signal hits the sensor solenoid, it starts resonating due to the frequency match. This resonance activates the internal controller and the clock. The clock signal is used to switch on and off the load resistor connected in parallel to the antenna. This is known as backscatter modulation. The reader antenna records the backscatter signal and based on the modulation it recognizes the tag. This technique is used in animal tracking.

References

- Website: https://www.mdpi.com/1424-8220/19/18/4012 accessed June 25, 2024.
- Website: https://rfid4u.com/inductive-and-backscatter-coupling/ accessed June 25, 2024.

Application 80: Satellite Beacon Signals

Satellite beacon signals are active Radio frequency signals used to identify the ground location and other geographic data to the satellites as shown in Fig. 6.11. Beacon signals are transmitted periodically at a fixed time interval. Using beacon signals the ground signals can receive important information from the satellites or aircraft by accurately tracking those. Since the satellites are in the far field, the power of the communication system is determined according to the plane wave propagation.

Fig. 6.11 Satellite
communication system

Application 81: Radio Spectrometers

Spectrometers analyze the power carried out by each electromagnetic frequency component. The spectrometers are heavily used in remote sensing. They can be categorized as active and passive spectrometers. Passive spectrometers rely on the natural electromagnetic spectrum to produce the signals while observing the received power. Once the power is received it breaks down the frequency components using a series of bandpass filters and analyzes the power carried by each frequency band. The active spectrometers transmit a certain range of frequencies and observe the reflected signal as shown in Fig. 6.12.

Radio telescopes are examples of spectrometers. They can be active or passive. The passive telescopes observe a region in space for a long period while collecting radio signals, then analyze the power in each frequency component. Active radio telescopes send out electromagnetic signals towards celestial bodies. Once these electromagnetic signals encounter a celestial body, they reflect from it. The signal reflects from the part of the celestial body closest to the Earth that arrives at the telescope first, and the signals further reflect from the further parts of the celestial body that arrive later. The processor connected to the telescope records the signal strength and the arrival time of the received signals. Once all these signals are analyzed and processed the shape of the celestial object and the distance from the Earth can be calculated. Also, considering the Doppler shift we can also determine whether this object is coming towards the Earth or going away from it. This technique is like the techniques used in radar.

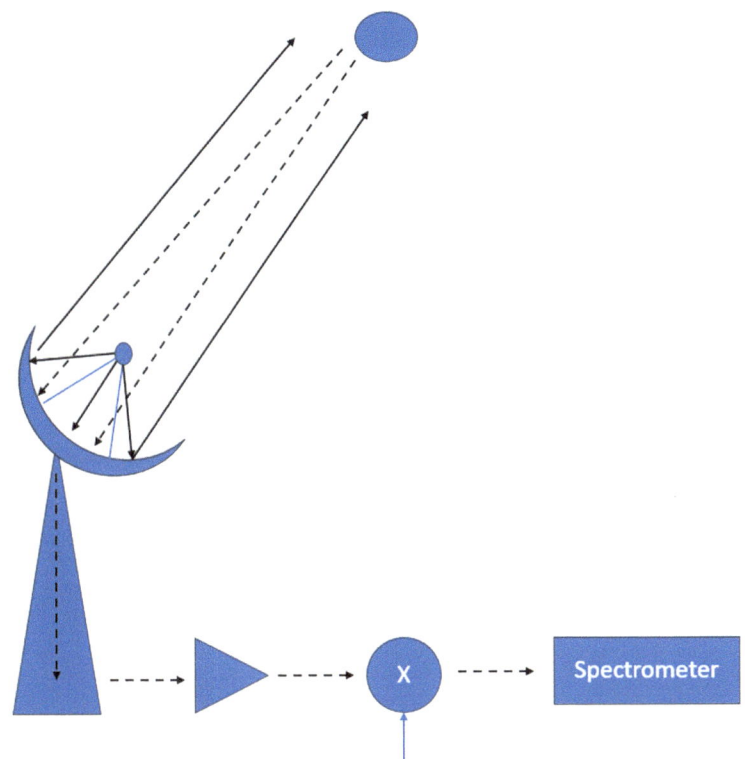

Fig. 6.12 Radio spectrometry

References

- Website: https://public.nrao.edu/telescopes/radio-telescopes/ accessed June 25, 2024.
- Website: https://www.planetary.org/articles/3248#:~:text=The%20radio%20dish%20r ecords%20the,wavelength%20on%20the%20other%20axis. Accessed June 25, 2024.

Application 82: Parallel Plate (Slab) Waveguide

When an electromagnetic field propagates in unbounded air in a vacuum it propagates in the speed of light. But, as we have experienced when the light is passed through a prism, it splits into multiple frequencies. This process is known as dispersion. The dispersion occurs when different frequencies travel at different velocities.

In a dispersive medium, there are two types of velocities: the group velocity and the phase velocity. In the example above, the white light consists of seven colors or frequencies. In this case, let's consider an electromagnetic wave composed of two frequencies:

ω_1 and ω_2.

$$s(z, t) = A\cos(\omega_1 t - k_1 z) + A\cos(\omega_2 t - k_2 z) \tag{6.13a}$$

$$s(z, t) = 2A\cos\left(\frac{\omega_1 - \omega_2}{2} t - \frac{k_1 - k_2}{2} z\right)\cos\left(\frac{\omega_1 + \omega_2}{2} t + \frac{k_1 + k_2}{2} z\right) \tag{6.13b}$$

This compound signal is similar to amplitude modulation, where the $\cos\left(\frac{\omega_1 - \omega_2}{2} t - \frac{k_1 - k_2}{2} z\right)$ represents the envelope or the slowly varying component of the compound signal and $\cos\left(\frac{\omega_1 + \omega_2}{2} t + \frac{k_1 + k_2}{2} z\right)$ represents the high-frequency carrier.

Let's consider these two components separately and get two expressions for the two velocities: one for the envelope and one for the phase of the carrier. The velocity of the envelope is called the group velocity since it shows the speed of a group of waves. This is also the speed at which energy travels. The high-frequency component represents the phase of each trough or the crest, hence called the phase velocity.

$$v_g = \frac{\omega_1 - \omega_2}{k_1 - k_2} = \frac{\Delta\omega}{\Delta k} \tag{6.13c}$$

$$\lim_{\Delta k \to 0} \frac{\Delta\omega}{\Delta k} = \frac{d\omega}{dk} \tag{6.13d}$$

$$v_p = \frac{\omega_1 + \omega_2}{k_1 + k_2} = \frac{\omega}{k} \tag{6.13e}$$

In free space, the group velocity and the phase velocity are the same. When the group and phase velocities are the same in a medium, it is called a non-dispersive medium. When the group and phase velocities are different in a medium, it is called a dispersive medium.

The group and phase velocities are important in studying electromagnetic wave propagation in waveguides, fiber optics, and plasma physics.

Let's consider the following notation for the electric field:

$$E = E_z e^{-\gamma z} e^{j\omega t} \tag{6.13f}$$

$$\mathbf{E_s} = \left(E_x \mathbf{a_x} + E_y \mathbf{a_y} + E_z \mathbf{a_z}\right) e^{-\gamma z} \tag{6.13g}$$

$$\mathbf{E} = \mathbf{E_s} e^{j\omega t} \tag{6.13h}$$

In the plane wave solution, we assumed,

$$\frac{\partial^2 \mathbf{E}}{\partial x^2} = \frac{\partial^2 \mathbf{E}}{\partial y^2} = 0 \tag{6.13i}$$

In the case of the waveguides, it is not the case. Hence, we need to consider the gradient operator in two parts.

$$\nabla^2 E = \nabla_t^2 E + \frac{\partial^2 E}{\partial z^2} \tag{6.13j}$$

$$\nabla^2 E = \mu\varepsilon \frac{\partial^2 E}{\partial t^2} \tag{6.13k}$$

Assuming time-harmonic variation,

$$\nabla_t^2 E + \frac{\partial^2 E}{\partial z^2} = (j\omega)^2 \mu\varepsilon E_s e^{j\omega t} \tag{6.13l}$$

Given that $e^{j\omega t}$, is redundant,

$$\nabla_t^2 E_s + \frac{\partial^2 E_s}{\partial z^2} = (j\omega)^2 \mu\varepsilon E_s \tag{6.13m}$$

Also, the variation of the spatial component concerning the direction of propagation z is $.e^{-\gamma z}$

Hence, the above expression becomes.

$$\nabla_t^2 E_s + \gamma^2 E_s = -k^2 E_s \tag{6.13n}$$

$k = \omega\sqrt{\mu\varepsilon}$ is called the wave number.

$$\nabla_t^2 E_s + = -(\gamma^2 + k^2)E_s \tag{6.13o}$$

Similarly for the magnetic field,

$$\nabla_t^2 H_s + = -(\gamma^2 + k^2)H_s \tag{6.13p}$$

The cutoff wave number is defined as;

$$k_c^2 = \left(\gamma^2 + k^2\right) \tag{6.13q}$$

Using the Faraday's Law, $\nabla \times E_s = -j\omega\mu H_s$

$$\frac{\partial E_z}{\partial y} + \gamma E_y = -j\omega\mu H_x \tag{6.13r}$$

$$-\frac{\partial E_z}{\partial x} - \gamma E_x = -j\omega\mu H_y \tag{6.13s}$$

$$\frac{\partial E_y}{\partial x} - \frac{\partial E_x}{\partial y} = -j\omega\mu H_z \tag{6.13t}$$

Using the Ampere's Law, $\nabla \times \boldsymbol{H_s} = j\omega\varepsilon\boldsymbol{E_s}$

$$\frac{\partial H_z}{\partial y} + \gamma H_y = j\omega\varepsilon E_x \tag{6.13u}$$

$$-\frac{\partial H_z}{\partial x} - \gamma H_x = j\omega\varepsilon E_y \tag{6.13v}$$

$$\frac{\partial H_y}{\partial x} - \frac{\partial H_x}{\partial y} = j\omega\varepsilon E_z \tag{6.13w}$$

$$E_x = -\frac{1}{k_c^2}\left(\gamma\frac{\partial E_z}{\partial x} + j\omega\mu\frac{\partial H_z}{\partial y}\right) \tag{6.13x}$$

$$H_x = \frac{1}{k_c^2}\left(-\gamma\frac{\partial H_z}{\partial x} + j\omega\varepsilon\frac{\partial E_z}{\partial y}\right) \tag{6.13y}$$

$$E_y = \frac{1}{k_c^2}\left(-\gamma\frac{\partial E_z}{\partial y} + j\omega\mu\frac{\partial H_z}{\partial x}\right) \tag{6.13z}$$

$$H_y = -\frac{1}{k_c^2}\left(\gamma\frac{\partial H_z}{\partial y} + j\omega\varepsilon\frac{\partial E_z}{\partial x}\right) \tag{6.13aa}$$

$$E_z = -\frac{j}{\omega\varepsilon}\left(\frac{\partial H_y}{\partial x} - \frac{\partial H_x}{\partial y}\right) \tag{6.13ab}$$

$$H_z = \frac{j}{\omega\mu}\left(\frac{\partial E_y}{\partial x} - \frac{\partial E_x}{\partial y}\right) \tag{6.13ac}$$

Let's consider Parallel Plate waveguide as shown in Fig. 6.13. In this case the electric and magnetic fields can propagate perpendicular to the direction of propagation. This is called the transverse electromagnetic (TEM) mode.

The following equations show the impedance, velocity inside the waveguide, conductor attenuation and the dielectric attenuation in the TEM mode.

$$Z_{TEM} = \frac{E_y}{H_x} = \eta = \eta_o\sqrt{\frac{\mu_r}{\varepsilon_r}} \tag{6.13ad}$$

Fig. 6.13 A parallel plate waveguide

b

$$v_{TEM} = \frac{1}{\sqrt{\mu\varepsilon}} = \frac{c}{\sqrt{\mu_r\varepsilon_r}} \tag{6.13ae}$$

$$\alpha_c = \frac{1}{\sigma\delta_c\eta b} \tag{6.13af}$$

$$\alpha_d = \frac{k^2 tan\delta_d}{2\beta} \tag{6.13ag}$$

Let's consider the scenario where the magnetic field is perpendicular to the direction of propagation. This is called the transverse magnetic or the TM mode $H_z = 0$

$$\frac{d^2E_z}{dy^2} = -k_c^2 E_z \tag{6.13ah}$$

$$E_z = \left[E_0 sin(k_c y) + E_1 cos(k_c y)\right]e^{-\gamma z} \tag{6.13ai}$$

The boundary condition to apply here is that, at $y = 0$ and $.y = b$, $E_z = 0$
Which will lead $E_1 = 0$ and $E_0 sin(k_c b) = 0$.
Hence $k_c = \frac{n\pi}{b}$; $n = 1, 2, 3 etc$

$$H_z = 0, E_z = 0 H_y = 0 \tag{6.13aj}$$

$$E_z = E_0 sin(k_c y)e^{-\gamma z} \tag{6.13ak}$$

$$E_y = -\frac{\gamma}{k_c^2}\frac{dE_z}{dy} = -\frac{\gamma}{k_c}E_0 cos(k_c y)e^{-\gamma z} \tag{6.13al}$$

$$H_x = \frac{j\omega\varepsilon}{k_c^2}\frac{dE_z}{dy} = \frac{j\omega\varepsilon}{k_c}E_0 cos(k_c y)e^{-\gamma z} \tag{6.13am}$$

Let's calculate the propagation constant.

$$\gamma = \sqrt{k_c^2 - k^2} = \sqrt{\left(\frac{n\pi}{b}\right)^2 - \omega^2\mu\varepsilon} \tag{6.13an}$$

The cutoff frequency is:

$$f_c = \frac{1}{2\pi}\frac{k_c}{\sqrt{\mu\varepsilon}} = \frac{1}{2\pi}\frac{n\pi}{b\sqrt{\mu\varepsilon}} = \frac{nv_{TEM}}{2b} \tag{6.13ao}$$

And the cutoff wavelength is:

$$\lambda_c = \frac{v_{TEM}}{f_c} = \frac{2b}{n} \tag{6.13ap}$$

The wavelength inside the waveguide is:

$$\lambda_g = \frac{2\pi}{\beta} = \frac{\lambda_{TEM}}{\sqrt{1 - \left(\frac{f_c}{f}\right)^2}} \tag{6.13aq}$$

And the phase velocity is:

$$v_p = \frac{\omega}{\beta} = \frac{v_{TEM}}{\sqrt{1 - \left(\frac{f_c}{f}\right)^2}} \tag{6.13ar}$$

The group velocity, impedance in the TM mode the conductor and dielectric attenuations for the TM mode are:

$$v_g = \frac{d\omega}{d\beta} = v_{TEM}\sqrt{1 - \left(\frac{f_c}{f}\right)^2} \tag{6.13as}$$

$$Z_{TM} = -\frac{E_y}{H_x} = \frac{\beta}{\omega\varepsilon} = \frac{\beta\eta}{k} \tag{6.13at}$$

$$\alpha_c = \frac{2k}{\sigma\delta_c\beta\eta b} \tag{6.13au}$$

$$\alpha_d = \frac{k^2 tan\delta_d}{2\beta} \tag{6.13av}$$

Now, let's consider the case, where the electric field is perpendicular to the direction of propagation known as the transverse electric or the TE Mode $E_z = 0$

$$E_z = 0, E_y = 0 H_x = 0 \tag{6.13aw}$$

$$H_z = H_0 \cos(k_c y)e^{-\gamma z} \tag{6.13ax}$$

$$E_x = \frac{j\omega\mu}{k_c}H_0 \sin(k_c y)e^{-\gamma z} \tag{6.13ay}$$

$$H_y = \frac{\gamma}{k_c}H_0 \sin(k_c y)e^{-\gamma z} \tag{6.13az}$$

The impedance for the TE mode is:

$$Z_{TE} = \frac{E_x}{H_y} = \frac{k\eta}{\beta} \tag{6.13ba}$$

The wavelength inside the waveguide, cutoff frequencies and all the other equations stay the same as the TM mode. The following two equations show the conductor and dielectric attenuations.

$$\alpha_c = \frac{2k_c^2}{\sigma \delta_c k \beta \eta b} \tag{6.13bb}$$

$$\alpha_d = \frac{k^2 tan\delta_d}{2\beta} \tag{6.13bc}$$

Reference

- Website: https://phys.libretexts.org/Bookshelves/Electricity_and_Magnetism/Electroma gnetics_II_(Ellingson)/06%3A_Waveguides/6.02%3A_Parallel_Plate_Waveguide-_Int roduction, accessed June 27, 2024.

Application 83: Rectangular Waveguide

Inside the rectangular waveguide shown in Fig. 6.14, the transverse electromagnetic mode is not possible.

Hence, for TM modes:

$$\nabla_t^2 E_z = \frac{\partial^2 E_z}{\partial x^2} + \frac{\partial^2 E_z}{\partial y^2} = -k_c^2 E_z \tag{6.14a}$$

$$E_z = Asin(k_x x)\sin(k_y y)e^{-\gamma z} \tag{6.14b}$$

$$k_c^2 = k_x^2 + k_y^2 \tag{6.14c}$$

$$k_x a = m\pi, m = 1,2, 3 \tag{6.14d}$$

$$k_y b = n\pi, n = 1,2, 3 \tag{6.14e}$$

Fig. 6.14 A rectangular waveguide

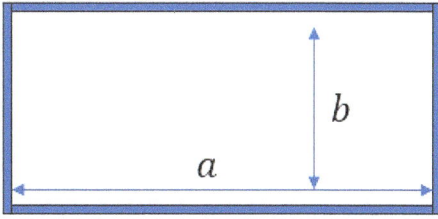

The cutoff frequency is:

$$f_{C_{mn}} = \frac{k_{C_{mn}}}{2\pi\sqrt{\mu\varepsilon}} = \frac{1}{2\pi\sqrt{\mu\varepsilon}}\sqrt{\left(\frac{m\pi}{a}\right)^2 + \left(\frac{n\pi}{b}\right)^2} \tag{6.14f}$$

$$E_x = -\frac{\gamma k_x}{k_{C_{mn}}^2}A\cos(k_x x)\sin(k_y y)e^{-\gamma z} \tag{6.14g}$$

$$H_x = \frac{j\omega\varepsilon k_y}{k_{C_{mn}}^2}A\sin(k_x x)\cos(k_y y)e^{-\gamma z} \tag{6.14h}$$

$$E_y = -\frac{\gamma k_y}{k_{C_{mn}}^2}A\sin(k_x x)\cos(k_y y)e^{-\gamma z} \tag{6.14i}$$

$$H_y = -\frac{j\omega\varepsilon k_x}{k_{C_{mn}}^2}A\cos(k_x x)\sin(k_y y)e^{-\gamma z} \tag{6.14j}$$

$$E_z = A\sin(k_x x)\sin(k_y y)e^{-\gamma z} \tag{6.14k}$$

$$H_z = 0$$

$$k_{C_{mn}}^2 = k_{x_m}^2 + k_{y_n}^2 = \left(\frac{m\pi}{a}\right)^2 + \left(\frac{n\pi}{b}\right)^2 \tag{6.14l}$$

The cutoff wavelength is:

$$\lambda_c = \frac{v_{TEM}}{f_{C_{mn}}} \tag{6.14m}$$

The wavelength inside the waveguide is:

$$\lambda_g = \frac{2\pi}{\beta} = \frac{\lambda_{TEM}}{\sqrt{1 - \left(\frac{f_{C_{mn}}}{f}\right)^2}} \tag{6.14n}$$

The phase velocity is:

$$v_p = \frac{\omega}{\beta} = \frac{v_{TEM}}{\sqrt{1 - \left(\frac{f_{C_{mn}}}{f}\right)^2}} \tag{6.14o}$$

The group velocity is:

$$v_g = \frac{d\omega}{d\beta} = v_{TEM}\sqrt{1 - \left(\frac{f_{C_{mn}}}{f}\right)^2} \tag{6.14p}$$

And the attenuation and the phase coefficients are:

$$\alpha = \sqrt{k_{C_{mn}}^2 - k^2} = k_{C_{mn}}\sqrt{1 - \left(\frac{f}{f_{C_{mn}}}\right)^2} \; ; f < f_{C_{mn}} \tag{6.14q}$$

$$\beta = \sqrt{k^2 - k_{C_{mn}}^2} = k\sqrt{1 - \left(\frac{f_{C_{mn}}}{f}\right)^2} \; ; f > f_{C_{mn}} \tag{6.14r}$$

For the transverse electric mode:

$$\nabla_t^2 H_z = \frac{\partial^2 H_z}{\partial x^2} + \frac{\partial^2 H_z}{\partial y^2} = -k_c^2 H_z \tag{6.14s}$$

$$E_x = \frac{j\omega\mu k_y}{k_{C_{mn}}^2} B cos(k_x x)\sin(k_y y)e^{-\gamma z} \tag{6.14t}$$

$$H_x = \frac{\gamma k_x}{k_{C_{mn}}^2} B sin(k_x x)\cos(k_y y)e^{-\gamma z} \tag{6.14u}$$

$$E_y = -\frac{j\omega\mu k_x}{k_{C_{mn}}^2} B sin(k_x x)\cos(k_y y)e^{-\gamma z} \tag{6.14v}$$

$$H_y = \frac{\gamma k_y}{k_{C_{mn}}^2} B cos(k_x x)\sin(k_y y)e^{-\gamma z} \tag{6.14w}$$

$$E_z = 0 \tag{6.14x}$$

$$H_z = B cos(k_x x)\cos(k_y y)e^{-\gamma z} \tag{6.14y}$$

Apart from the above the cutoff frequency, wavelength and the other parameters are the same for TE and TM modes.

Reference

- Website: https://phys.libretexts.org/Bookshelves/Electricity_and_Magnetism/Electroma gnetics_II_(Ellingson)/06%3A_Waveguides/6.08%3A_Rectangular_Waveguide-_TM_ Modes, accessed June 27, 2024.

Application 84: Magnetron

Magnetron shown in Fig. 6.15 is a device that is used to convert high-power voltages and currents into microwaves. In other words, these are microwave sources. Let's start the discussion of the magnetron with the feed. The magnetron contains a capacitor and

Fig. 6.15 Magnetron

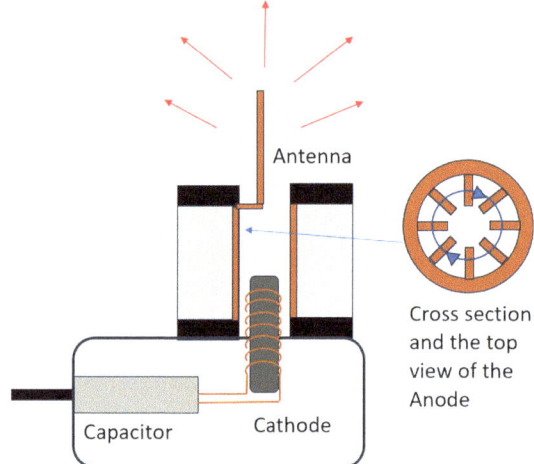

Antenna

Cross section
and the top
view of the
Anode

Capacitor   Cathode

an inductor (cathode) connected in series. This capacitor-inductor combination generates *LC* oscillations. The inductor or the cathode releases electrons by thermionic emissions.

The next step is to accelerate these electrons. That is done using an anode. With the placement of the anode, the electrons emitted by the cathode are accelerated towards the anode. To increase the time these electrons, spend inside the anode cavity two magnets are used at the top and the bottom of the anode cavity. These magnets force the electrons to take a curved path increasing their time spent inside the anode cavity. Also, the anode is designed such that each cavity wall has an induced potential hence the electric field pattern looks like a solenoid. As the capacitor and the cathode go through oscillations the electrons released from the cathode also oscillate, making the induced voltage on the anode oscillate as well. The anode is connected to an antenna to convert these voltage oscillations into electromagnetic waves. The frequency of these waves is controlled by controlling the *LC* oscillation frequency.

References

- Website: https://www.youtube.com/watch?v=bUsS5KUMLvw accessed June 25, 2024.
- Website: https://www.cpii.com/docs/related/2/MAG%20TECH%20ART.pdf accessed June 25, 2024.

Application 85: Electromagnetic Interference and Electromagnetic Compatibility

Given the relationship between the electromagnetic waves, voltage, and current the electromagnetic energy can be coupled into different systems. This coupling can be radiative, inductive, conductive, or capacitive. Radiative coupling occurs when the radiating electromagnetic waves interact with material such as in radiography. Exposure to radiative electromagnetic energy for a long period can damage tissues and cause bodily organ failure. Inductive coupling occurs when there is a linkage of magnetic flux. In transformers, this inductive coupling is used productively. But when this coupling is undesirable it causes interference.

Conductive coupling occurs when there is charge flow from one point to another. The capacitive coupling occurs due to the potential difference between two points.

Electromagnetic compatibility is the ability of electromagnetic instruments to operate in their usual electromagnetic environments. For example, all electromagnetic instruments are equipped with electromagnetic noise cancellation and filtering techniques. Electromagnetic compatibility standards must be followed by the equipment developers regardless of the brand. For example, a mobile phone should be able to connect with the base station transceiver regardless of the make.

Practice Problems

1. Let's consider the communication between a cell phone and the tower. Cell phones use small-scale smart antennas that can identify the direction of the tower. Therefore, power is directed toward the direction of the tower, instead of transmitting in all directions. In the United States cell phone communication uses 850 MHz frequency.
   a. Cell phone antenna radiation power is 3 W, and the total power is distributed on a surface area of 1/16 (one-sixteenth) of a sphere based on its location. If the direction between the cell phone and the tower is 1km (1000 m), calculate the magnitude of average power density received at the tower receiving antenna. The medium of propagation is free space.
   b. Using your answer in part (a), calculate the magnitude of the electric field intensity received at the tower antenna.
   c. For protection tower receiving antenna is coated with polythene which has a relative permittivity of 2.25, relative permeability of 1, and conductivity of 0.001 S/m. Determine whether polythene is a low-loss dielectric or a conductor at 850 MHz frequency.
   d. Based on your answer for part (c) calculate the attenuation coefficient for polythene.
   e. The electric field intensity you obtained in part (b) is at the coated outer surface of the receiving antenna. If the polythene coating is 1 mm, calculate the magnitude of electric field intensity measured by the inner antenna circuitry after attenuation by the polythene coating.

Fig. 6.16 The radiator and its cover

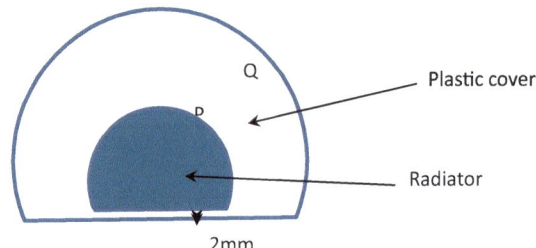

2. Average radiated power and Poynting Theorem.

A 5 GHz Wi-Fi access point (network router) mounted on the ceiling inside a room with length, width, and height dimensions 20 m × 20 m × 7 m. The transmit power of the router is 200mW. A person keeps a laptop on a table with a height of 1 m, and a mobile phone right underneath the tabletop. The width of the tabletop is 3 cm

a. Calculate the wavelength of the emitted EM signal.

b. Since the room dimensions are much larger than the wavelength of the signal, we can assume that the radiation pattern is half of a sphere. How much is the received average EM power density by the laptop?

c. The tabletop is made with melamine, which has an electrical conductivity of 0.00102 S and a relative permittivity of 4.7. Determine whether melamine acts as a low-loss dielectric or a conductor at 5 GHz.

d. Based on your answer in part c, calculate the attenuation constant for melamine.

e. How much would be the magnitude of the average power density received by the mobile phone? For now, you may ignore the phase angle introduced by the impedance of melamine.

3. An electromagnetic radiator is at the center of a soccer field is shown in Fig. 6.16.

The radiation pattern of the radiator can be considered as half a sphere. The radiation power of the radiator is 2 kW and the transmission frequency is 1 GHz. A student wants to measure the electromagnetic power of the radiator. The radial distance between the radiator and the student's handheld power meter is 20 m.

a. Considering total free space propagation find the average power density measured by the handheld power meter.

b. How much of an electric field amplitude is available at the location of the student? Assume free space propagation.

c. To protect against humidity, the student is going to cover the radiator with a plastic cover that has a thickness of 2 mm. Plastic roughly has a relative permittivity of 2.0 and conductivity of 0.01S/m. The relative permeability of plastic is 1. Calculate the attenuation constant α for plastic at 1 GHz.

d. If an electric field intensity amplitude of 1228 V/m, is measured at point P on the radiator, how much of an electric field intensity amplitude can be measured at point Q on the outer surface of the plastic cover?

e. Beyond the plastic cover, the medium is free space. How much of an electric field intensity amplitude can be measured at the location of the student after covering the radiator? Assume the location of the student did not change.

4. Let's consider the communication link between the battleship Washington and the submarine USS Neptune.

Battleship Washington is on the surface of the ocean while USS Neptune is 500m below sea level. Figure 6.16 shows the battleship and the submarine. The horizontal distance between the ship and the submarine is 1km. The battleship Washington transmits its signals at a frequency of 30 kHz, with an output power of 25 watts. The battleship intends to communicate with other ships; hence the radiation is confined to a half–a–sphere in the air. In the diagram, point P is in the air, and point Q is in seawater right underneath point P.

a. Considering free space propagation and the horizontal distance between the ship and the submarine, calculate the power density at point P, directly above the submarine.

b. Hence calculate the magnitude of the electric field intensity at point P.

c. If the conductivity(sigma) and the relative permittivity of seawater are 4.3 S/m and 80 respectively, determine whether seawater acts as a low-loss dielectric or a conductor at 30 kHz.

d. Assuming the relative permeability of seawater to be 1, calculate the attenuation constant of seawater and the impedance.

e. When an electromagnetic wave encounters a boundary of media only a portion of the wave is being transmitted while the other is being reflected. Calculate the transmission coefficient for the waves at the air-sea water boundary. Make sure to express the transmission coefficient in magnitude and phase form.

f. Remember the phase of the impedance indicates a phase offset in the transmitted signal concerning the incident signal. Given that, and the answer in part b, calculate the magnitude of the electric field intensity at point Q.

g. Now, using the attenuation constant calculated in part d, and the electric field intensity at point Q calculate the magnitude of the electric field intensity at the submarine.

h. Based on your answer to part g, give a reason, why 30 kHz is not a good frequency to communicate with submarines.

This chapter presents the applications of Maxwell's equations in testing materials. Material properties play an important role in the semiconductor industry as well as in the aerospace and space industries.

Application 86: Capacitive Measurement of Complex Permittivity

Permittivity is the ability of a material to resist an external electric field by creating an internal electric field. So far, we considered the permittivity as a real quantity. But, in some applications, the permittivity is expressed as a complex quantity. Permittivity is studied extensively in the areas of materials and electronic fabrication. For example, when selecting dielectrics for capacitors and when selecting substrates for high-performance printed circuit boards. In addition, with the expansion of biomedical applications, the dielectric industry's importance increased.

In this section, let's study the complex permittivity in detail.

The Ampere's Law with the Maxwell's contribution says:

$$\nabla \times \boldsymbol{H} = \boldsymbol{J} + \frac{\partial \boldsymbol{D}}{\partial t} \tag{7.1}$$

Assuming time harmonic electric fields, $\boldsymbol{E} = \boldsymbol{E}_s e^{j\omega t}$, where $\boldsymbol{E}_s$ is the spatial—vector component.

$$\nabla \times \boldsymbol{H}_s = \sigma \boldsymbol{E}_s + j\omega\varepsilon \boldsymbol{E}_s \tag{7.1a}$$

$$\nabla \times \boldsymbol{H}_s = j\omega\left(\varepsilon - \frac{j\sigma}{\omega}\right)\boldsymbol{E}_s \tag{7.1b}$$

For materials with non-zero conductivity, the imaginary portion of the permittivity is defined as:

$$\varepsilon'' = \frac{\sigma}{\omega} \tag{7.1c}$$

And the real part of permittivity given in Eq. 7.1d is the permittivity that we have used so far.

$$\varepsilon' = \varepsilon \tag{7.1d}$$

The electric loss tangent defined above is based on the tangent of the imaginary and real parts.

$$\tan\delta = \frac{\varepsilon''}{\varepsilon''} = \frac{\sigma}{\omega\varepsilon} \tag{7.1e}$$

However, there is another aspect that we need to consider when it comes to complex permittivity. That is the relationship between the electric field intensity and the electric flux density.

The electric field intensity $\boldsymbol{E}$ and the electric flux density $\boldsymbol{D}$ are related by the permittivity.

$$\boldsymbol{D} = \left(\varepsilon' - j\varepsilon''\right)\boldsymbol{E} \tag{7.1f}$$

When there is a time-varying electric field, the electric flux generation is not instantaneous. Hence the relationship between $\boldsymbol{E}$ and $\boldsymbol{D}$ becomes:

$$\boldsymbol{D} = a_o\boldsymbol{E} + a_1\frac{\partial \boldsymbol{E}}{\partial t} + a_2\frac{\partial^2 \boldsymbol{E}}{\partial t^2} + a_3\frac{\partial^3 \boldsymbol{E}}{\partial t^3}\ldots. \tag{7.1g}$$

This relationship shows that for both the real and imaginary parts of the permittivity, there are contributions from the higher-order derivatives of the time-harmonic electric field intensity. Previous research has shown that up to 1 GHz, the real part of the permittivity stays constant, and the imaginary part changes according to $\frac{\sigma}{\omega}$.

These frequency variations are used in material testing. We saw in a previous section that; different materials behave differently at different frequencies. Based on the observation that at low frequencies (<1 GHz) the real part of the permittivity stays constant, and the imaginary part is due to the conductivity of the material, the permittivity of a material can be measured using impedance testing or capacitive testing. In this setup, the material under test (MUT) is sandwiched between two metal electrodes, and the impedance of the capacitor is measured. Figure 7.1 shows the test set-up.

Fig. 7.1 The complex permittivity measurement set-up

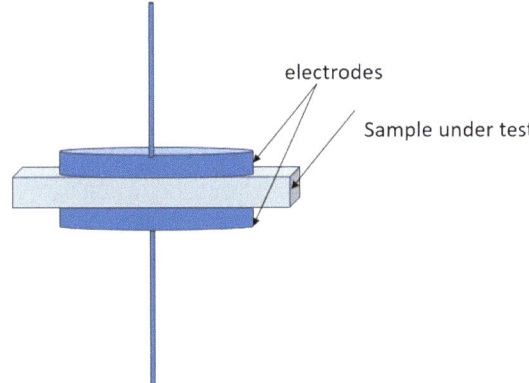

The capacitance is given as:

$$C = \frac{\varepsilon_r \varepsilon_o S}{d} \tag{7.1h}$$

In Eq. 7.1 h, S Üitor plate (electrode in this case), and d is the thickness of the dielectric. Let's remind ourselves that the real part of the permittivity is:

$$\varepsilon' = \varepsilon_r \varepsilon_o \tag{7.1i}$$

The measurement steps are as follows:

I. input voltage and measure the output current.
II. calculate the impedance $Z = \frac{v}{i}$.
III. Calculate the capacitance $C = |Z|\omega$
IV. Calculate the real-relative permittivity: $\varepsilon'_r = \frac{Cd}{\varepsilon_o S}$
V. Measure the loss tangent—The imaginary part of the complex permittivity is due to the conductive losses at low frequencies. In other words, at low frequencies, the capacitor deviates from its ideal behavior which can be modeled by a capacitor in series with a resistor known as the series equivalence resistance.

The impedance of an ideal capacitor is purely reactive. The loss tangent is the tangent of the angle between the impedance of the capacitor in the complex plane and the negative reactive axis.

VI. Calculate the imaginary- relative permittivity: $\varepsilon''_r = \varepsilon'_r tan\delta$
VII. Express the full complex—relative permittivity $\varepsilon'_r - j\varepsilon''_r$
VIII. Express the full complex permittivity: $\varepsilon'_r \varepsilon_o - j\varepsilon''_r \varepsilon_o$

References

- Website: https://phys.libretexts.org/Bookshelves/Electricity_and_Magnetism/Electroma gnetics_II_(Ellingson)/03%3A_Wave_Propagation_in_General_Media/3.04%3A_Com plex_Permittivity accessed June 25, 2024.
- Website: https://en.wikipedia.org/wiki/Dissipation_factor#:~:text=A%20real%20capa citor%20has%20a,and%20the%20negative%20reactive%20axis.&text=is%20often% 20expressed%20as%20a%20percentage) accessed June 25, 2024.

Application 87: Calculating the Dissipation Factor in Budgeting Signal Loss

As a signal propagates through a transmission line, it attenuates. This attenuation is calculated as follows in the units of Decibels per inch.

$$Attenuation\left(\frac{dB}{in}\right) = 2.3 \times f \times \tan\delta \times \sqrt{\varepsilon_r} \qquad (7.2)$$

Consider a very high-frequency circuit design running at 1 GHz, made with typical FR4 material. Which has a relative permittivity of 4 and the attenuation at 1 GHz is 0.1 dB/in. If the trace length of this on the PCB is 30 inches, a 3 dB loss needs to be budgeted into the design at the beginning. Otherwise, the receiver will be receiving a signal 3 dB lower than expected. Reference

https://www.intel.com/content/www/us/en/docs/programmable/683883/current/loss-tan gent.html#:~:text=Loss%20tangent%20(tan(%CE%B4)),the%20dissipation%20factor% 20(Df) accessed June 25, 2024.

Application 88: Skin Depth and Its Importance

Skin depth is defined as the reciprocal of the attenuation constant. This is a measurement of how deep an electromagnetic field can penetrate a material before being attenuated. If a material is conductive at a given frequency, then it absorbs energy from the electromagnetic field to mobilize its electrons. This process is what causes the attenuation.

$$skin\,depth(\delta) = \frac{1}{\alpha} \qquad (7.3)$$

Skin depth plays a major role in applications such as biomedical engineering and geoscience among many other fields that use non-invasive or minimally invasive techniques. For example, non-invasive blood glucose spectroscopy requires a signal to penetrate below the skin adipose (fat) tissues of the skin, and geological surveys require an electromagnetic signal to penetrate the ground to discover different constituents such as water and minerals. Skin depth is measured in meters.

Depending on the skin depth or the penetration depth required, the transmission frequency should be adjusted, such that the transmission frequency should be lower for a higher skin depth and higher for a lower skin depth.

Reference

- Website: https://www.epa.gov/environmental-geophysics/electromagnetic-signal-attenuation-skin-depth accessed June 25, 2024.

Application 89: Magnetization Measurement Using the Extraction Method

The simplest way to measure the magnetization is called the extraction method. In this case, the sample is kept inside a solenoid, within another solenoid called the search coil as shown in Fig. 7.2. The magnetic flux is measured with and without the sample.

$$\phi_{m1} = \mu_o(\boldsymbol{H} + \boldsymbol{M}) \tag{7.4a}$$

Without the sample:

$$\phi_{m2} = \mu_o\boldsymbol{H} \tag{7.4b}$$

From the Faraday's law, the induced EMF in both cases:

Fig. 7.2 Extraction method measurement set-up

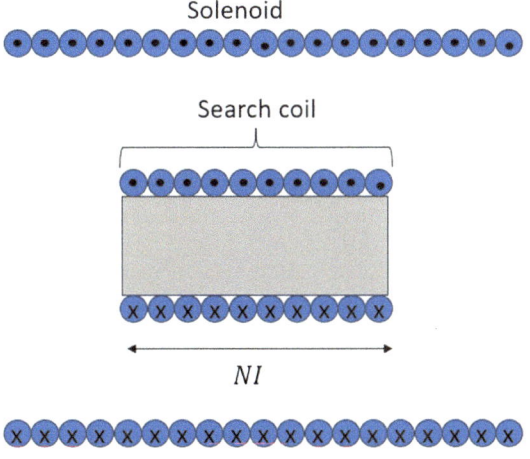

$$V_1(t) = -N \frac{\partial \phi_{m1}}{\partial t} \rightarrow \phi_{m1} = -\frac{1}{N} \int V_1(t).dt \qquad (7.4c)$$

Similarly.

$$V_2(t) = -N \frac{\partial \phi_{m2}}{\partial t} \rightarrow \phi_{m2} = -\frac{1}{N} \int V_2(t).dt \qquad (7.4d)$$

N is the number of turns within the search coil. Based on the above two equations the magnetization M can be calculated.

Application 90: Measuring the Magnetization of a Material—Vibration Sample Magnetometry

In vibration sample magnetometry, the magnetization is measured by vibrating the sample perpendicular to the magnetic field as shown in Fig. 7.3. From the Faraday's Law:

$$\oint_l E.dl = -\frac{\partial \phi_m}{\partial t} \qquad (7.5)$$

The voltage or the electromotive force induced in a pickup coil is:

Fig. 7.3 Vibrating sample magnetometry

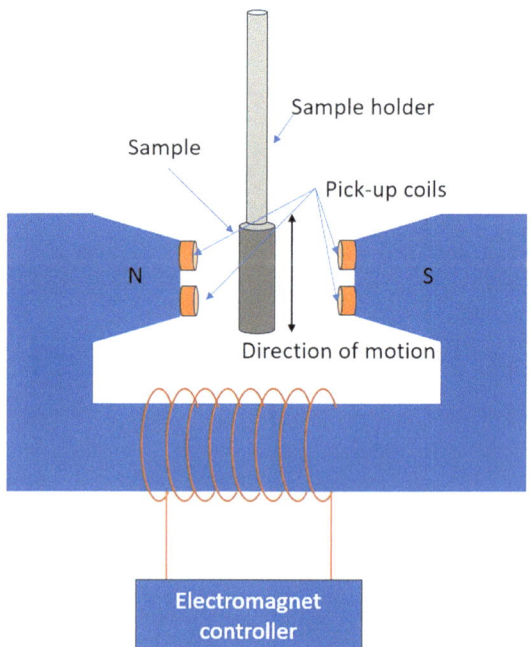

$$V = -\frac{\partial \phi_m}{\partial t} = -\frac{\partial \phi_m}{\partial z} \cdot \frac{\partial z}{\partial t} \tag{7.5a}$$

The $\frac{\partial z}{\partial t}$ term indicates the vertical vibration of the sample. This motion is time-harmonic with the frequency f_{vib}. Therefore:

$$z(t) = sin(2\pi f_{vib}t) \tag{7.5b}$$

$$\frac{\partial z}{\partial t} = 2\pi f_{vib}cos(2\pi f_{vib}t) \tag{7.5c}$$

Now, let's see what the $\frac{\partial \phi_m}{\partial z}$ value is. So far, we know the magnetization as:

$$\chi_m = \frac{M}{H} \rightarrow M = \chi_m H \tag{7.5d}$$

If the expand the simple linear system given in Eq. 7.5d and generalized it to a linear time invariant system of an isotropic medium, we get the Eq. 7.5e.

$$M(r, t) = \chi_m(r, t) * H(r, t) \tag{7.5e}$$

In Eq. 7.5e, $*$ is the convolution operation, r is any position vector and t is time. The Fourier transform of Eq. 7.5e gives the relationship in Eq. 7.5f in the frequency domain.

$$M(r, \omega) = \chi_m(r, \omega)H(r, \omega) \tag{7.5f}$$

The Eq. 7.5e is valid in isotropic media. The vibrating sample magnetometry can measure the magnetic properties of anisotropic media as well. In anisotropic media, the magnetic susceptibility becomes a tensor. A sample tensor in Cartesian coordinates is given in Eq. 7.5f.

$$\overleftrightarrow{\chi_m(r,\omega)} = \begin{pmatrix} \chi_m(x, \omega) \\ \chi_m(y, \omega) \\ \chi_m(z, \omega) \end{pmatrix} \tag{7.5g}$$

Like the definition that the electric polarization is the electric dipole moment within a unit volume, magnetization is the magnetic dipole moment per unit volume. Let the magnetic dipole moment be m.

This magnetic dipole moment is created by magnetic poles (hypothetical) within a displacement of l.

Based on the above scenario:

$$M = \frac{m}{v} \tag{7.5h}$$

If the total sample shifts with each oscillation.

$$\frac{\partial \phi_m}{\partial z} = \mu_o MA \tag{7.5i}$$

In Eq. 7.5i, A is the area of each coil. This equation can be expressed in terms of the magnetic dipole moment $\boldsymbol{m}$:

$$\frac{\partial \phi_m}{\partial z} = \mu_o \boldsymbol{m} v A \tag{7.5j}$$

In Eq. 7.5j, v is the volume of the sample. The combined vA product can be called g, a geometrical factor.

The induced voltage picked up by a coil is:

$$V = -2\pi f_{vib}\mu_o \boldsymbol{m} g \cos(2\pi f_{vib}t) \tag{7.5k}$$

Given that the volume of the sample, the cross-sectional area of the coil, and the oscillation frequency are known, the only unknown, the magnetic dipole moment can be calculated using the induced voltage equation.

Reference

- Website: https://etheses.dur.ac.uk/5506/1/5506_2945.PDF accessed June 25, 2024.

Application 91: Radiography (X-ray Imaging)

X-rays are highly energetic electromagnetic waves. The frequency of the X-rays is ranging from 30 petahertz to 30 exahertz and the wavelengths range from 0.01 to 10 nm. Given the short wavelength, these are called short waves. The X-rays were discovered by Wilhelm Rontgen (1845–1923) and have been a prominent discovery in medical imaging.

To explain the operational principles of radiography we need to consider the particle behavior of electromagnetic waves known as photons. A photon is considered an energy packet. The energy of photons increases with the frequency according to the following equation.

$$E_{photon} = hf \tag{7.6}$$

In the above equation, h is the Plank's constant which has the exact value of $6.62607015 \times 10^{-34} JHz^{-1}$.

The X-rays are generated inside an X-ray tube as shown in Fig. 7.4. Inside the X-ray tube the cathode current releases electrons by thermal excitation. These electrons are accelerated towards the anode and upon hitting the anode release their energy-producing bremsstrahlung.

The maximum energy of these bremsstrahlung photons can be calculated as:

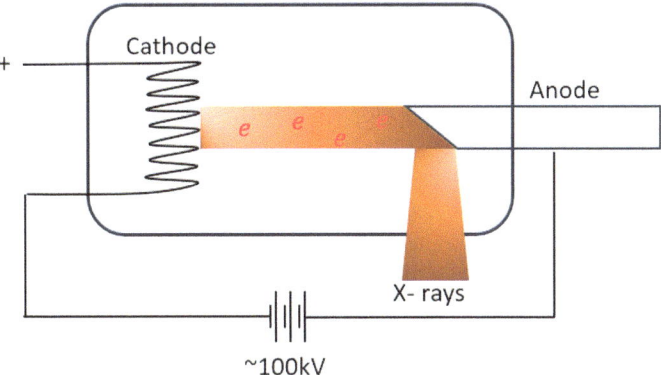

Fig. 7.4 The X-ray tube

$$E_{max} = eV \tag{7.6a}$$

In Eq. 7.6a, e is the charge of an electron and V is the X-ray tube voltage.

$$E_{max} = 1.602 \times 10^{-19} \text{ C.100 kV} \tag{7.6b}$$

$$E_{max} = 1.602 \times 10^{-19}.100 \text{ kJ} = 100 \text{ keV} \tag{7.6c}$$

Note that the unit of energy eV is equal to

$$1.602 \times 10^{-19} \text{ J} \tag{7.6d}$$

X-ray bremsstrahlung produces a continuous X-ray spectrum. There is another radiation mechanism known as the characteristic radiation. Characteristic radiation occurs when an electron bombards the anode it can release an electron from an orbital shell (ex: K-shell). This leaves a space known as a hole. To fill this hole, another electron from the higher energy outer shell (L-shell) drops into the hole releasing the energy difference between the two shells $E_{L-shell} - E_{K-shell}$. The characteristic radiation looks like a narrow peak superimposed on the continuous bremsstrahlung spectrum. In an X-ray tube the number of photons is controlled by the cathode current (6–100 mA) and the energy of photons is controlled by the cathode voltage (50–125 kV).

Multiple types of interactions may occur when the X-ray beam is exposed to matter:

i. Photoelectric absorption- photoelectric absorption occurs when the X-ray photons are absorbed by an atom. The energy absorption releases an electron, and this released electron is ejected in the same direction as the incoming photon is traveling.

ii. Compton scattering—in Compton scattering the photon transfers only part of its energy to an electron. This energy reduction in the photon makes it deviate from the direction of incoming photons, and the electron that absorbed part of the energy is ejected in a different direction.

iii. Pair production—in pair production, the photon is transformed into an electron and a positron (with equal mass to an electron but with opposite charge). This positron will recombine with another electron and create two photons that escape in opposite directions. For this mechanism to take place the initial photon should have an energy level of 1 MeV or higher.

iv. The X-ray photons can initiate nuclear reactions inside matter. Although such high energies are not used in medical imaging, this is what prevents prolonged exposure to X-rays.

Photoelectric absorption dominates in lower energies and at intermediate energies Compton scattering dominates. Also, with the increasing atomic number of the material, the photoelectric absorption increases more rapidly than Compton scattering. Pair production occurs only at very high energies.

In X-ray imaging what is being imaged is the intensity of the output beam (I_{out}). When an X-ray beam travels through material with thickness $d = x_{out} - x_{in}$, the difference between the input (I_{in}) and output beam intensity for a single beam is given by:

$$I_{out} = I_{in}\exp\left(-\int_{x_{in}}^{x_{out}} \alpha(x)dx\right) \tag{7.6e}$$

In Eq. 7.6e, $\alpha(x)$ is the location-dependent attenuation coefficient of the material.

Reference

- Suetens P. *Fundamentals of Medical Imaging*. 2nd ed. Cambridge University Press; 2009.

Application 92: Transmission Electron Microscopy

Transmission electron microscopy is a high-resolution imaging technique to image samples at an micro and nanometer scales. In transmission electron microscopy or TEM, electrons are launched from a tungsten electron gun. These electrons are accelerated towards the anode as shown in Fig. 7.5. This electron beam travels through the vacuum column at the center of the microscope. The TEM uses magnetic lenses to converge the electron beam. The magnetic lenses consist of coils and pole pieces and uses the difference in the Lorentz force to converge the electron beam to a focal point. As the electron beam hits the sample suspended in the vacuum column, the interaction between the electrons and the atoms in the sample scatter the electron beam creating an image similar to an X-ray image on the florescent screen.

Fig. 7.5 Components of a
transmission electron
microscope (TEM)

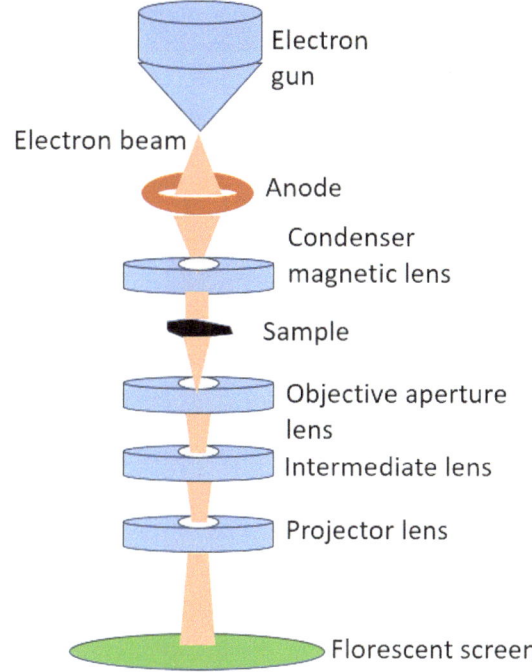

Reference

- Website: https://microbenotes.com/transmission-electron-microscope-tem/#:~:text=
  To%20study%20and%20differentiate%20between,on%20compounds%20and%20t
  heir%20structures., accessed July 31, 2024.

Application 93: Scanning Electron Microscopy

Figure 7.6 shows a schematic of a scanning electron microscope (SEM). Similar to the transmission electron microscope, the scanning electron microscope also uses an electron gun and an anode. This instrument also uses magnetic lenses to focus the electron beam. The only difference being the placement of the sample. Similar to the transmission electron microscope, the scanning electron microscope also images the interaction between the electron beam with the sample. In scanning electron microscopy, the reflected electron (backscattered electrons) and the characteristic X-ray emissions, and the newly generated electrons (secondary electrons) are measured to generate an image. Once a particular location was imaged the focal point is moved to another location to produce a 2D image.

Fig. 7.6 Components of a scanning electron microscope (SEM)

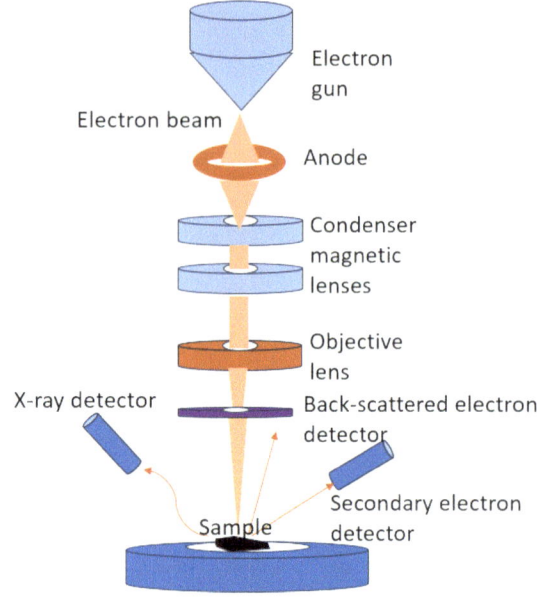

Reference

- Website: https://www.nanoscience.com/techniques/scanning-electron-microscopy/#:~:
  text=The%20position%20of%20the%20electron,then%20detected%20by%20appropr
  iate%20detectors. Accessed July 31, 2024.

Application 94: Nuclear Magnetic Resonance

Nuclear magnetic resonance is based not only on the fundamental laws of electromagnetics but also on the laws of quantum physics. Nuclear magnetic resonance uses the magnetic properties of the nucleus. The nucleus of an atom acts as a small magnet. These magnetic spin around its axis like a top. This motion is called precession. Under normal circumstances, these spins are randomly oriented. When a sample of these nuclei is exposed to an external magnetic field, these nuclei orient themselves either parallel or antiparallel to the external magnetic field. The energy of parallel ($E_\uparrow$) and antiparallel ($E_\downarrow$) states are described in quantum physics as follows:

$$E_\uparrow = -\frac{1}{2}\gamma\hbar B_o \tag{7.7a}$$

$$E_\downarrow = +\frac{1}{2}\gamma\hbar B_o \tag{7.7b}$$

In the above equations, $-\frac{1}{2}$ and $+\frac{1}{2}$ are the spin quantum numbers. $\hbar$ is the reduced Plank's constant ($\frac{h}{2\pi}$), γ is the gyromagnetic ratio which is a constant for a particular nucleus and B_o is the field strength of the external magnetic field.

The nuclei processing anti-parallel to the external magnetic field are at a higher energy state than the nuclei processing parallel to the magnetic field. The resonance condition of the nuclei is defined as:

$$E_\downarrow - E_\uparrow = \gamma \hbar B_o = \hbar \omega_o \tag{7.7c}$$

The angular frequency $\omega_o = \gamma B_o$ is known as the Larmour frequency. For protons, the Larmor frequency is 42.6 MHz at $B_o = 1T$. The Larmor frequency differs with the atomic number.

The NMR process is as follows:

i. A sample is kept under the main magnetic field. Which makes the nuclei precess either parallel or anti-parallel to the main magnetic field.
ii. An RF pulse with a frequency equal to the Larmor frequency is applied to the sample. This will make some of the nuclei parallel to the magnetic field to achieve the anti-parallel energy state. This will create a resulting magnetic field perpendicular to the main magnetic field.
iii. Once the RF pulse is removed the nuclei relax back to their original energy state. This is known as free induction decay. As the nuclei relax back to their original energy state, the change in their radii induces a voltage (EMF) on the detector coil as shown in Fig. 7.7.
iv. This detected time-domain signal is then transformed into the frequency domain using Fourier transform. Hence the resonance conditions of different particles are identified.

References

- Suetens P. *Fundamentals of Medical Imaging*. 2nd ed. Cambridge University Press; 2009.
- Website: https://www.journalofyoungphysicists.org/post/the-principles-and-applicati ons-of-magnetic-resonance-a-comprehensive-review-of-mri-technology#:~:text=Far aday's%20law%20determines%20that%20a,denoted%20by%20a%20magnetic%20m oment accessed June 25, 2024.

Fig. 7.7 The nuclear magnetic
resonance set-up

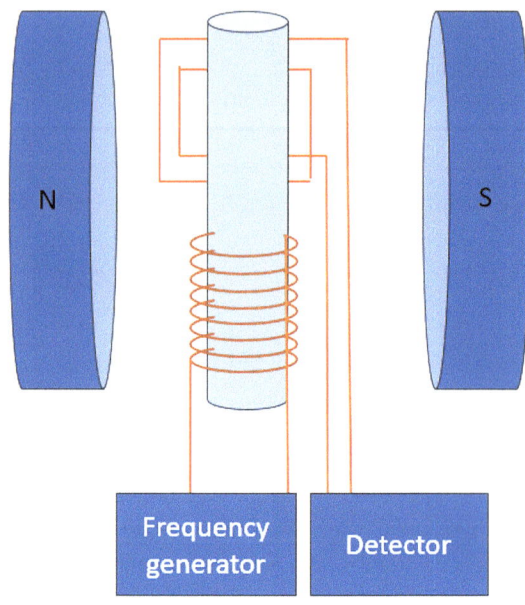

Application 95: Magnetic Resonance Imaging (MRI)

The Magnetic Resonance Imaging uses the principle of nuclear magnetic resonance. The MRI machine is comprised of a main magnet that stays always on, an RF magnetic coil to produce the RF pulse, and the gradient magnetic field to achieve more spatial resolution as shown in Fig. 7.8. Like the technique mentioned in the NMR section here, once a patient is lying on the table, the nuclei of the tissues align with the main magnetic field, creating parallel and antiparallel processes. According to the quantum theory, a little over half of the nuclei process parallel to the magnetic field while the rest process anti-parallel. Once the RF pulse of Larmor frequency is applied on the patient, some low energy state particles achieve high energy states creating a transverse net magnetic field. Due to this transverse net magnetic field, this initial RF pulse is known as the 90° pulse.

Once the 90° pulse is removed, the nuclei achieve their original energy state through free induction decay. The free induction decay induces a voltage or a current signal which is measured by the detector. The time taken to achieve 67% of the original magnetization in the longitudinal direction is called the T1 relaxation time, and the time taken to reduce the transverse magnetization by 67% is called the T2 relaxation time.

The free induction decay occurs only after the 90° pulse and it's a rapid process. To prolong this process to get a measurable signal the 90° is followed by a 180° pulse applied along the direction of the main magnetic field. The purpose of the 180° pulse is to phase synchronize the nuclei. The signal produced after the 180° pulse is known as the echo.

Fig. 7.8 Coil arrangement of a
MRI machine

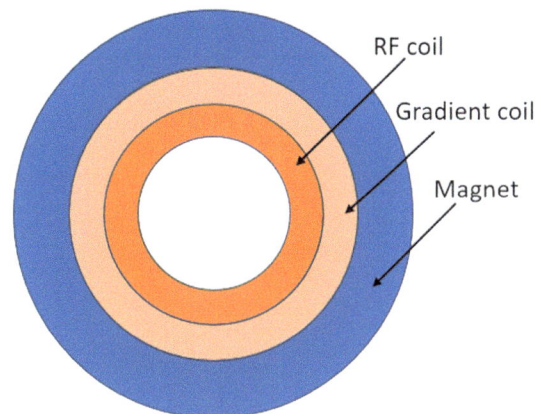

The combination of the 90° pulse followed by the 180° pulse is called the spin-echo sequence.

References

- Website: https://www.youtube.com/watch?v=jLnuPKhKXVM accessed June 25, 2024.
- Website: https://www.barrowneuro.org/for-physicians-researchers/education/grand-rou nds-publications-media/barrow-quarterly/volume-16-no-2-2000/basics-of-magnetic-resonance-imaging/#:~:text=The%20production%20of%20electricity%20by,image% 20of%20in%20vivo%20magnetism. Accessed June 25, 2024.
- Website: https://www.nibib.nih.gov/science-education/science-topics/magnetic-res onance-imaging-mri#:~:text=MRI%20of%20a%20knee.,pull%20of%20the%20magn etic%20field. Accessed June 25, 2024.
- Website: https://mriquestions.com/gradient-coils.html#:~:text=Gradients%20are%20l oops%20of%20wire,secondary%20magnetic%20field%20is%20created. Accessed June 25, 2024.

Application 96: Microwave Oven

A microwave oven is the best example of electromagnetic radiation. All major components of a microwave oven are related to Maxwell's equations. Figure 7.9 the major components of a microwave oven. First, let's start with the high-voltage transformer. As we know already, the working principle of the transformers is Faraday's Law, more specifically known as the transformer EMF, where an AC voltage is used to induce an electromotive force in the secondary coil. Inside a microwave, a step-up transformer is used to amplify the voltage. The output of the transformer is fed to the magnetron. Inside the magnetron,

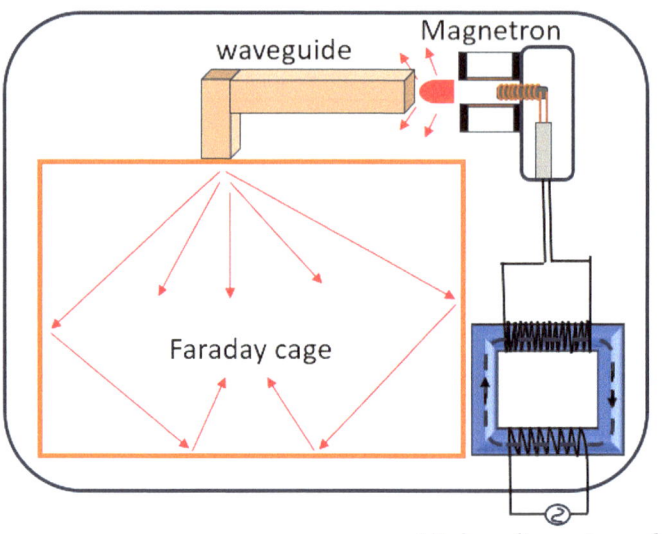

Fig. 7.9 The major components of a microwave oven

the voltage and current are transformed into microwaves. These are called microwaves since the wavelengths of the waves are in the micrometer range. The magnetron is coupled to a waveguide. The waveguide carries the waves into the microwave heating chamber which is a Faraday cage. The purpose of this Faraday cage is to contain the microwave energy inside the chamber. This is also called a resonant cavity. When matter such as food is placed inside the microwave, the microwave energy is converted into thermal energy.

A newer technique is called microwave-assisted induction heating (MAIH). In MAIH, in addition to typical microwave heating, induction heating is used at the bottom of the oven to expedite the heating process.

Reference

- Website: https://www.intechopen.com/chapters/76372 accessed June 25, 2024.

Application 97: Polarization Measurement of a Dielectric Material

In the electric dipole example, the product term qd is called the electric dipole moment $\boldsymbol{p}$. In vector form, this is the product of electric charge and the displacement $\boldsymbol{d}$. The electric dipole moment is a measurement of positive and negative charge separation within a system. Previously we learned that the electric polarization is:

$$P = \chi_e \varepsilon_o E \tag{7.8a}$$

In Eq. 7.8a, χ_e is the electric susceptibility which indicates the degree of polarization. Another definition for electric polarization is the volumetric dipole moments in a dielectric material.

$$P = \frac{p}{v} \tag{7.8b}$$

The charges within a dielectric material are usually at equilibrium. When disturbed by an electric field these charges are being displaced by a distance d creating the dipole moment. Since the charges in dielectrics are not free to move as in conductors these are called bound charges. The bound charge density inside a dielectric is:

$$\rho_b = -\nabla . P \tag{7.8c}$$

The negative sign is because the net flux out is considered positive according to Gauss's law hence the flux inward of the dielectric is considered positive. The polarization is measured using a simple capacitive technique as follows:

$$|P| = \frac{qd}{Sd} \tag{7.8d}$$

$$|P| = \frac{CV}{S} \tag{7.8e}$$

In the equation above, C is the capacitance, V is the voltage across the capacitor and S is the surface area of a capacitor plate.

References

- Website: https://interestingengineering.com/science/what-is-electric-polarization-and-how-does-it-affect-our-daily-lives accessed June 25, 2024.
- Website: https://warwick.ac.uk/fac/cross_fac/xmas/xmasbeamline/xmas_offline/ele ctrical_measurements/#:~:text=The%20polarization%20is%20calculated%20simp ly,area%20of%20the%20sample%20electrodes. Accessed June 25, 2024.
- Website: https://en.wikipedia.org/wiki/Polarization_density accessed June 25, 2024.

Application 98: Frequency Domain Induced Polarization and Complex Resistivity

Induced polarization and complex resistivity are simple, but a powerful technique used in geophysics to determine the permittivity of soil. In this technique four probes are used: two to measure current and two to measure the potential difference as shown in Fig. 7.10.

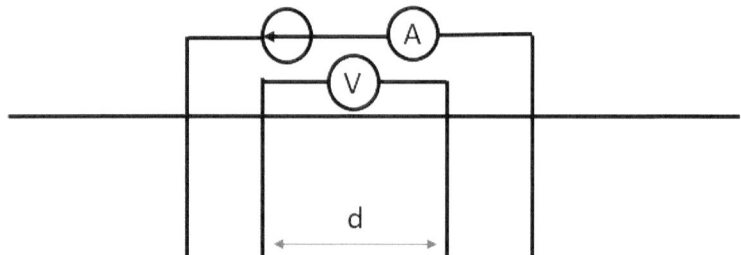

Fig. 7.10 Voltage and current electrode placement in measuring complex resistivity

While taking measurements, both the resistance and capacitance measurements are taken simultaneously. In time domain-induced polarization analysis, a direct current is injected into the material (soil). Whereas in frequency domain analysis, two alternating currents with two frequencies are injected.

Soil contains multiple constituents and some of these are polarized atoms. Under normal conditions, these constituents are in quasi-equilibrium. When disturbed by an external electric field, these atoms arrange themselves to shield the external electric field. The charge movement in shielding this electric field creates a capacitance. Based on the constituents in the soil the conductivity and the resistivity changes. But as we know already, the resistance is frequency independent. While the capacitance is dependent on frequency. The capacitance is a measurement of the permittivity of the material while the resistance indicates the conductivity.

Let's start with the equation of continuation of current:

$$J = \sigma E \tag{7.9}$$

$$\sigma = \frac{1}{\rho} \tag{7.9a}$$

In Eq. 7.9a, ρ is the resistivity.

For the above system, let's consider two frequencies ω_f, and ω_s, which represent fast and slow frequencies.

The equation for the system becomes:

$$I_o e^{j\omega_f t} = \left(\frac{1}{R}V_f e^{j\omega_f t} + C_f \frac{d}{dt}V_f e^{j\omega_f t}\right) = \left(\frac{1}{R} + jC_f\omega_f\right)V_f e^{j\omega_f t} \tag{7.9b}$$

$$I_o e^{j\omega_s t} = \left(\frac{1}{R}V_s e^{j\omega_s t} + C_f \frac{d}{dt}V_s e^{j\omega_s t}\right) = \left(\frac{1}{R} + jC_s\omega_s\right)V_s e^{j\omega_s t} \tag{7.9c}$$

The complex conductance in the fast and slow cases are:

$$G_f^* = \left(\frac{1}{R} + jC_f\omega_f \right) \tag{7.9d}$$

$$G_s^* = \left(\frac{1}{R} + jC_s\omega_s \right) \tag{7.9e}$$

The magnitudes of both the currents and voltage can be measured. Also, the phase angles between the voltage and the current are:

$$\varphi_f = \tan^{-1}\left(\frac{C_f\omega_f}{R} \right) \tag{7.9f}$$

$$\varphi_s = \tan^{-1}\left(\frac{C_s\omega_s}{R} \right) \tag{7.9g}$$

The resistance is the same for both frequencies and the frequencies are known. These capacitors can be treated as parallel plate capacitors. Hence the relative permittivity of soil can be calculated using:

$$C = \frac{\varepsilon_r\varepsilon_o S}{d} \tag{7.9h}$$

In Eq. 7.9h, S is the surface area of an electrode and d is the distance between the potential difference probes.

References

- Website: https://www.britannica.com/science/electric-polarization accessed June 25, 2024.
- Website: https://www.epa.gov/environmental-geophysics/induced-polarization-ip-and-complex-resistivity accessed June 25, 2024.

Application 99: Maxwell's Equations in Bio-medical Applications

This section is a brief induction to the application of Maxwell's equations in biomedical applications. Radiation therapy has been used in tissue abnormality detection, cancer treatments as well as in non-invasive measurements.

In external beam radiotherapy high energy electromagnetic beam is focused into the region of the body where the cancerous tissues are present. In noninvasive impedance spectroscopy, the changes of the impedance of the tissues are measured to detect the abnormalities. In this technique, the sensors used are microstrip patch antennas.

Another area is the modeling of tissues. Depending on the composition the permittivity of the human tissues is different. Hence the tissues of the human body are modeled using the complex permittivity. These complex permittivity measurements are combined with the nonlinear Schrodinger's equation to model the responses of the human tissues with plane waves.

Reference

- Website: https://www.ncbi.nlm.nih.gov/pmc/articles/PMC4437579/ accessed June 25, 2024.

Application 100: Magneto Hydrodynamics

Magnetohydrodynamics or MHD is a powerful analysis tool in fluid dynamics and physics. This powerful tool is widely applied in analyzing the space plasma such as the magnetic flux from the sun. Magnetohydrodynamics connects the Maxwell's equations with the governing equations of fluid dynamics. The parameters we consider here are the plasma mass density ρ, the magnetic field B, the kinetic pressure P, and the plasma velocity v.It is worth noting that in magnetohydrodynamics plasma is considered as a fluid. The 1st equation of MHD is on the continuity and the conservation of mass:

$$\frac{\partial \rho}{\partial t} + \nabla(\rho v) = 0 \tag{7.10a}$$

The second equation is the equation of motion of an element of the fluid. This is also called the Euler's equation.

$$\rho \left[\frac{\partial v}{\partial t} + (v\nabla)v \right] = -\nabla P + J \times B \tag{7.10b}$$

In the above equation, J is the electric current density.

The 3rd equation is the equation of energy. In its simplest adiabatic case:

$$\frac{d}{dt}\left(\frac{P}{\rho^\gamma} \right) = 0 \tag{7.10c}$$

γ is the ratio of specific heats and normally this value is 5/3.

$$P = \sum_{i=1}^{n} \frac{2k_B}{m_i} \rho T \tag{7.10d}$$

k_B is the Boltzmann's constant.

To derive the fourth MHD equation, let's start with the Lorentz force:

$$F = q(E + v \times B) \tag{7.10e}$$

When there are no external forces:

$$E = -v \times B \qquad (7.10\text{f})$$

Now, let's apply the Faradays law of induction:

$$\nabla \times E = -\frac{\partial B}{\partial t} \qquad (7.10\text{g})$$

This brings us to the fourth MHD equation on the magnetic dynamo or the MDH induction equation:

$$\frac{\partial B}{\partial t} = \nabla \times (v \times B) \qquad (7.10\text{h})$$

Practice Problems

1. Consider an electric field given as:

$$E(x, y, z) = 8.12a_y \frac{V}{m}$$

   a. For the above electrically polarized material, the electric flux density is calculated to be;

$$D(x, y, z) = 0.2876a_y \frac{nC}{m^2}$$

   Calculate the relative permittivity of the dielectric medium.
   b. How much is the susceptibility (χ) of the above electrically polarized material?
   c. Hence calculate the polarization of the above material.
   d. We are going to change the permittivity of this electrically polarized material such that, it becomes un-polarized. What should be the new permittivity of this unpolarized material?
2. Although an MRI machine contains a large coil to produce the magnetic field, let's consider the operation of a single loop. Assume that the direction of the magnetic field is $-a_x$.
   a. In the MRI machine, the magnetic flux density (B) is increased from 0 to 2T (Tesla) within 0.25 seconds. Hence, calculate the rate of change of magnetic flux density.
   b. If the radius of the coil is 50 cm, how much is the rate of change of magnetic flux through the loop?
   c. Using Faraday's law determine the magnitude of the electric field intensity generated within the loop.
   d. Based on Stokes's theorem, comment on the curl of the electric field intensity generated within the loop (the value of it, and is the field rotational or irrotational?)

Hint: think about the point form of Faraday's law.
3. Polarization of diamonds and graphite.

Here, you will be analyzing the polarization characteristics of diamonds and graphite; the two forms of carbon crystal structures. The dielectric constant of diamond is 5.7 and for graphite, it is 15.

a. Calculate the electrical susceptibility of diamond and graphite.
b. If a piece of diamond and a piece of graphite are exposed to an electric field E of $0.5a_x$ V/m, calculate of polarization for the diamond and graphite.
c. How much would be the electric flux density for both diamond and graphite?
d. Diamond has a stronger crystal structure compared to graphite. If we keep the electric field at 0.5 V/m for graphite and change the electric field across the diamond, how much should be the new electric field magnitude to achieve the same polarization for both elements?

4. Commercial flat-panel TVs have Barium-Silicon Monoxide as the polarizing agent which has a relative permittivity of 8.

a. How much is the electric susceptibility of the material?
b. If an electric field of 0.5 mV/m was applied on the above material, calculate the magnitude of the polarization.
c. How much would be the magnitude of the electric flux density?
d. If the surface area of the TV is 133 cm × 75 cm how much is the total electric charge on the TV screen?
e. How much is the surface charge density on the TV screen?